P9-ELW-098

© John Connor Press Associates

About the Author

MITCHELL SYMONS is the author of *That Book . . . of Perfectly Useless Information*, *This Book . . . of More Perfectly Useless Information*, and *Why Girls Can't Throw . . . and Other Questions You Always Wanted Answered*. He worked at BBC TV as a researcher and then a director before leaving to become a full-time writer. A columnist for the *Sunday Express*, Symons writes for numerous publications. He lives on the southern coast of England.

Where Do Nudists Keep Their Hankies?

Also by Mitchell Symons

FICTION

All In

The Lot

NONFICTION

Forfeit!

The Equation Book of Sports Crosswords

The Equation Book of Movie Crosswords

The You *Magazine Book of Journolists* (four books, coauthor)

Movielists (coauthor)

The Sunday Magazine *Book of Crosswords*

The Hello! *Magazine Book of Crosswords* (three books)

How to Be Fat: The Chip and Fry Diet (coauthor)

The Book of Criminal Records

The Book of Lists

The Book of Celebrity Lists

The Book of Celebrity Sex Lists

The Bill Clinton Joke Book

National Lottery Big Draw 2000 (coauthor)

That Book

This Book

The Other Book

The Sudoku Institute (co-author)

Why Girls Can't Throw

How to Speak Celebrity

Where Do Nudists Keep Their Hankies?

. . . and Other Naughty Questions You Always Wanted Answered

Mitchell Symons

HARPERENTERTAINMENT

NEW YORK • LONDON • TORONTO • SYDNEY

HARPER**ENTERTAINMENT**

WHERE DO NUDISTS KEEP THEIR HANKIES?. Copyright © 2007 by Mitchell Symons. All rights reserved. Printed in the United States of America. No part of this book may be used or reproduced in any manner whatsoever without written permission, except in the case of brief quotations embodied in critical articles and reviews. For information address HarperCollins Publishers, 10 East 53rd Street, New York, NY 10022.

HarperCollins books may be purchased for educational, business, or sales promotional use. For information please write: Special Markets Department, HarperCollins Publishers, 10 East 53rd Street, New York, NY 10022.

FIRST EDITION

Designed by Joy O'Meara

Library of Congress Cataloging-in-Publication Data is available upon request.

ISBN: 978-0-06-113407-4

ISBN-10: 0-06-113407-4

07 08 09 10 11 ID/RRD 10 9 8 7 6 5 4 3 2 1

To Penny, Jack, and Charlie
with all my love

Where Do Nudists Keep Their Hankies?

Introduction

"Is sex dirty?" asked Woody Allen before answering his own question: "Only if it's done right." Well, I've never *shtupped* my ex-girlfriend's adopted daughter (not much of a boast), so I guess I'll have to bow to Mr. Allen's greater knowledge, but let me pose a rhetorical question of my own: Is sex funny? Indeed, is there a more hilarious sight than a man before, during, or after the act of sex?

And part of what makes sex so comical is the fact that people are so damn serious about it.

As a happily married man, I'm more than happy to watch from the sidelines—so to speak—making snide and unhelpful comments at all the ridiculous participants.

Anything for a chuckle.

Of course, I've got experience in this area. Although my last book (*Why Girls Can't Throw*) wasn't specifically about sex, I managed to sneak in lots of sexual questions—such as:

If you were supple enough to give yourself a blow job, would that make you gay?

Why are lesbians called dykes?

All right, are those rumors about Richard Gere and gerbils true?

Is it true that Catherine the Great died having sex with her horse?

Is it true that Marianne Faithfull once had a foursome with Mick Jagger, Keith Richards, and a Mars bar?

Do women who live/work together menstruate at the same time?

What is the origin of the use of the word "gay" to mean "homosexual"?

Who was the best-endowed celebrity of all time?

Has anyone ever died during sex?

Do women have wet dreams too?

Well, now I've decided to devote a whole book to questions about sex.

Of course, I'm just a middle-aged man who's convinced that everyone's having more fun than me. One of the more unfortunate aspects of the human condition is the fear that you're missing out. It was ever thus. As Michael Caine said of the 1960s, "Swinging London was the same two hundred people in the Kings Road sleeping with each other." I was never one of the two hundred—except maybe for a two-month period in 1981, when every girl seemed to say yes and . . . well, I don't expect you to accompany me on that particular trip down memory lane. In any event, other people's sex lives are more interesting than your own. Even if, in truth, they're not.

Clearly, I'm no sex expert, but then who is? I suppose "sex experts" are, but they do take sex so bloody seriously. I had a run-in with these people once, which illustrates this perfectly. It happened about twenty years ago. I used to write regular features for *Penthouse*. They were a bit "naughty" but also really funny—in many ways, the best things I ever wrote. Alas, all my contributions will have been lost: I shouldn't think any copies have survived the purges that soon-to-be-attached young men are obliged to go through—hence the piles of ripped-up jazz mags you find on rubbish heaps.

Anyway, I got a call from the then-editor of *Forum*, the bible for the sexually serious, asking me to go through all the ads from the past year and, from them, put together a sexual map of Britain. A job's a job; and, as a freelancer (then as now), I only had one question: How much?

So, I started scanning the small ads and I was soon astonished by how much I didn't know. There were perversions—"preferences," as the editor corrected me—that were just unbelievable. One was infibulation, which, I gathered through a fug of terror, involved your loved one putting a tiny padlock through your genital area. Well, not my loved one—at least not while I had any strength in my body.

Anyway, I wrote the piece as well as I could but I couldn't resist spiking it with gags. I recall writing about infibulation that while I respected the diversity of other people's sex lives, I had to report a nightmare I'd had in which I was late with my copy and the editor came round with a giant padlock . . .

It didn't go down well. Nor, in fact, did any of my bons mots. When I protested to the editor that sex was funny and should be treated as such, she explained to me that *Forum* readers took it seriously—very seriously. They were used to sexologists who also took sex seriously—very seriously.

I never wrote for the magazine again and have had contempt for sexologists—amateur and professional—ever since. However, that's not to say that I'm without backup in this area. My friends and acquaintances include several gays and lesbians, two men who became women, one woman who became a man, several adulterers, a satyr, a former prostitute, a celibate, and a couple who have an "open marriage." I also have two teenage sons—although I've promised them I'll only consult them in an emergency. I also have access to doctors, zoologists, philosophers, and psychiatrists.

In case you think that I'm kind of louche or anything, I should say that I doubt that my friends and acquaintances are any more outré or sophisticated than yours: my guess is that your circle contains a remarkably similar mix, it's just that you've never delved that deep.

Fortunately, as my friends have often had cause to remark, I lack both the tact and the embarrassment genes and so am totally uninhibited about asking incredibly personal questions. I'm driven by two specific things: a very low threshold of boredom and almost insatiable curiosity. It's a toxic mix that sometimes makes my wife wince with embarrassment (she, alas for her, has the e. gene) and occasionally loses us friends but, at long last, I can put it to some practical use.

So, here are a hundred or so questions of a sexual nature: breezy not sleazy; entertaining not titillating—if any bloke reading this gets a hard-on then I will have failed. (If, on the other hand, any woman—any extremely cute woman—finds herself getting incredibly aroused by something I've written, then my e-mail address can be found at the back of the book. I am, of course, happily married and totally monogamous but, hey, everything's negotiable . . .)

You Like to Do *What*?

(Or, How Perverted Is Perverted?)

"The only unnatural sex act is that which you cannot perform."

(ALFRED KINSEY)

"The only unnatural sexual behavior is none at all."

(SIGMUND FREUD)

In some ways, this is the signature question for the whole book, in the sense that it covers the waterfront. My first port of call was good old Google. I put in "sexual practices," but after tiptoeing smartly around the sort of sites you really don't want your wife to find in your browser, I went offline and consulted the books in my really rather impressive library (where size counts) of sex books—especially *The Encyclopaedia of Unusual Sexual Practices*—and found a series of activities ranging from the hey-that's-me-I-didn't-know-it-had-a-name! to the sounds-fun to the not-for-me-but-I-can-get-my-head-around-it to the why-would-anyone-want-to-do-that/how-did-anyone-even-think-of-that to the oh-my-god-just-don't-go-there.

To give you an idea, let me give you examples from each

category (omitting any "preferences" that will be the subject of separate questions).

Hey-That's-Me-I-Didn't-Know-It-Had-a-Name!

Allorgasmia: fantasizing about someone other than one's partner
Basoexia: being aroused by kissing
Gymnophilia: being aroused by nudity
Mammagymnophilia: being aroused by breasts

Sounds-Fun

Antholagnia: being aroused by smelling flowers
Axillism: rubbing of penis in an armpit
Bigynist: sex between one male and two females (if only!)
Coitus à mammilla: rubbing of penis between breasts
Coitus à unda: sex or sex games in water
Tripsophilia: being aroused by massage

Not-For-Me-But-I-Can-Get-My-Head-Around-It

Acousticophilia: being aroused by sounds
Acrophilia: being aroused by heights or high altitudes
Agonophilia: person who is aroused by partner pretending to struggle
Agrexophilia: being aroused by others knowing you are having sex
Albutophilia: being aroused by water
Amaurophilia: preference for a blind or blindfolded sex partner
Anasteemaphilia: attraction to taller or shorter partners

Androminetophilia: being aroused by female partner who dresses like a male

Amomaxia: sex in parked car

Autopederasty: person inserting their own penis into their anus

Capnolagnia: being aroused by watching others smoke

Dacryphilia: person who is aroused by seeing their partner cry

Dendrophilia: being aroused by tree or fertility worship of them

Dogging: couples who engage in sex in their car while others watch from outside

Doraphilia: being aroused by animal fur, leather, or skin

Erotographomania: being aroused by writing love poems or letters

Frottage: rubbing body against partner or object for arousal

Genuphallation: insertion of penis between the knees of a partner

Gynemimetophilia: being aroused by a male impersonating a female

Hirsutophilia: being aroused by armpit hair

Hodophilia: being aroused by traveling

Knismolagnia: being aroused by tickling

Lactaphilia: being aroused by lactating breasts

Lygerastia: tendency to be aroused only in darkness

Maieusiophilia: being aroused by pregnant women

Melolagnia: being aroused by music

Moriaphilia: being aroused by telling sexual jokes

Nasophilia: nose fetish

Neophilia: being aroused by novelty or change

Oculophilia: eye fetish

Odontophilia: being aroused by teeth
Olfactophilia: being aroused by smells
Podophilia: foot fetish
Pygophilia: being aroused by contact with buttocks
Trichophilia: hair fetish
Xenophilia: being aroused by strangers
Zelophilia: being aroused by jealousy

Why-Would-Anyone-Want-to-Do-That/ How-Did-Anyone-Even-Think-of-That

Acrotomophílía: sexual preference for amputees
Agalmatophilia: being aroused by statues
Algophilia: being aroused by experiencing pain
Androidism: being aroused by robots with human features
Apotemnophilia: person who has sexual fantasies about losing a limb
Arachnephilia: being aroused by spiders
Asphyxiaphilia: being aroused by lack of oxygen
Asthenolagnia: being aroused by weakness or being humiliated
Autassassinophilia: being aroused by orchestrating one's own death by the hands of another
Ball dancing: self-flagellation by hanging fruit from hooks in skin
Blood sports: sex games that involve blood
Belonephilia: being aroused by use of needles
Catheterophilia: being aroused by use of catheters
Chezolagnia: masturbating while defecating
Coprophilia: being aroused by playing with feces
Entomophilia: being aroused by insects
Eproctophilia: being aroused by flatulence

Formicophilia: sex play with ants
Harmatophilia: being aroused by sexual incompetence or mistakes, usually in female partner
Harpaxophilia: being aroused by being robbed or burgled
Hemotigolagnia: being aroused by bloody sanitary pads
Hierophilia: being aroused by sacred objects
Homilophilia: being aroused by hearing or giving sermons
Iantronudia: being aroused by exposing oneself to a doctor
Idrophrodisia: being aroused by BO, especially from the genitals
Kleptophilia: being aroused by stealing
Klismaphilia: being aroused by enemas
Menophilia: being aroused by menstruating women
Mysophilia: being aroused by dirt
Ophidiophilia: being aroused by snakes
Pecattiphilia: being aroused by sinning or possibly guilt
Pediophilia: being aroused by dolls
Peodeiktophilia: exhibitionism
Phobophilia: being aroused by fear or hate
Phygephilia: being aroused by being a fugitive
Psychrophilia: arousal from being cold or watching others freeze
Pyrophilia: being aroused by fire or its use in sex play
Scopophilia: being aroused by getting stared at
Siderodromophilia: being aroused by trains
Taphephilia: being aroused by getting buried alive

Oh-My-God-Just-Don't-Go-There

Autophagy: self-cannibalism or eating one's own flesh
Autosadism: infliction of pain or injury on oneself
Brachioprotic eroticism: a deep form of fisting where the arm enters the anus

Dysmorphophilia: being aroused by deformed or physically impaired partners
Emetophilia: being aroused by vomit
Nosophilia: being aroused by knowing partner has terminal illness
Symphorophilia: being aroused by arranging a disaster, crash, or explosion
Anything involving children, incest, violence, and degradation. Oh yes, and necrophilia—sex with corpses.

OK, so that's how I rate those different "preferences," but my reactions are necessarily idiosyncratic. What I consider fun or reasonable might shock you, and (almost certainly) vice versa. I guess the bottom line is that whatever consenting adults consent to do in private is entirely their business. But note the importance of the three operative words: consenting (and that consent must carry with it the wherewithal to consent), adults, and private.

After putting together the above list, I e-mailed it to my friend Rick, whom I've already trailed as a satyr (dictionary definition: "a man with strong sexual desires"—yup, that's Rick). He phoned me and said, "You prig!" (I might have misheard him). "Just because you're a boring little fart who has sex three times a week and then only in the missionary position, how DARE you judge what other people get up to? For your information, you prick"—I heard him right that time—"I have done several of the things that you say 'aren't for you,' a few of the things that you're too fucking timid to have even considered—or so you say—and, yes, there are one or two things on your oh-my-god-just-don't-go-there-because-I'm-such-a-fucking-prude list that I'd be prepared to consider because, unlike you, I'm open-minded!"

Such as?

"Go fuck yourself!"

I was about to say "I would but I'm not an autopederast" (see above), but he'd hung up, leaving me yelping, "I'm not a prude!"—the four words which, as I reflected later, do more than any others to mark out . . . yes, the prude.

What Do Men Really Want?

I canvassed all my male friends and acquaintances on this one and, perhaps not surprisingly, the consensus was as follows:

1. To get it and . . .
2. . . . to get away with it.

"Many a husband kisses with his eyes wide-open. He wants to make sure his wife is not around to catch him."

(ANTHONY QUINN)

"I wasn't kissing her, I was whispering in her mouth."

(CHICO MARX WHEN HIS WIFE CAUGHT HIM KISSING ANOTHER WOMAN)

◎ In order to mate, a male deep sea anglerfish will bite a female when he finds her. The male will never let go and will eventually merge his body into the female and spend the rest of his life inside the female mate. The male's internal organs will disappear except the testes, which are needed for breeding.

What Do Women Really Want?

That's me in the spotlight, resisting the temptation to say, "Who cares?" As a man, I start from the assumption that men want sex whereas women want . . . well, so much more.

Now I know I'm generalizing—no, *really*?—but there's something in this, you know, and it's more, much more than the fact that women need foreplay and affection and bloody telephone calls afterwards whereas men just want a quick shag.

There is, I suspect, a different biological imperative at work here.

To confirm my suspicion, I went to visit my pet social anthropologist, Dr. Lorraine Mackintosh, which was no hardship, as she is, as I have remarked in my previous books, no pig to look at.

Is there, I asked, a biological imperative at work here?

"That's the trouble with you hacks: you get hold of an expression—like biological imperative—and you start tossing it around like a, like a . . ."

Salad?

"You know what I mean. In fact, you're not entirely wrong. There *is* a biological imperative at work here. Irrespective of the fact that twenty-first-century women act differently from their cavewomen ancestors, a surprising number of atavistic fears and drives persist. Chief among these is to procure the seed of the alpha male and then to find a male—not necessarily the same one—to help nurture that offspring. The man, on the other hand, just wants to spread his seed as far and as wide as possible."

So, a bloke wants to screw around whereas a bird wants the full works?

"Er, not quite."

OK, can we say that, for men, having sex is the end—the aim, if you like—but that for women, it's just the beginning?

"No, not really. Yes, biologically, men want to impregnate as many women as possible, but women don't necessarily want to hang on to their men forever. This is also an explanation of why some women—many of us, if we're honest—are attracted to bastards. Do you remember how I said that women didn't necessarily need the same man to impregnate them and then provide for the child?"

Yes.

"Well, that's terribly important to bear in mind. The woman needs the very best seed and that's likely to belong to the alpha male who is, in turn, likely to be a bit of a bastard. Having got herself impregnated by this man, she now has her work cut out to keep him because *his* instinct is to go off and spread his seed. The trouble is that the very qualities she was searching for in a biological father for her child are unlikely to coincide with the qualities she'll need from the man who will support her and this child. Hence the conflict between the bastard and the nice guy."

So, how is this conflict resolved?

"Good question. Very often it's not. The girl finds her bastard but, before long, he's off screwing around. Alternatively, she has her good old Mr. Dependable who's always there for her, but she finds him dull and boring and so she starts to stray . . ."

What, in search of a bastard?

"Precisely," said Dr. M, rewarding me with a dazzling smile.

So, we're doomed then, are we, as a species?

"No. Obviously not! The alpha male can learn to suppress his wandering and apply his superiority over his fellow males to the task of making lots of money for his family. After all, it's in the alpha male's interests to produce successful offspring—"

—and sticking around to raise them is probably a better bet than screwing around—or, as we experts call it, the scattergun approach.

"There's a fascinating thing you should know about the male. What's the first thing his mother-in-law says after his wife—her daughter—has given birth? 'Doesn't he look like his father?' That's right, the mother-in-law's role is to help solidify the father's bond to his child and the best way to do that is to persuade him that it is indeed his child."

Fascinating.

"As for the sturdy, dependable type—the antithesis, if you like, to the alpha male—his relationship with his wife can thrive if she can project onto him the sort of alpha-male qualities she yearns for."

Which explains why the most unlikely blokes are touted as tigers by their wives. But does this help to explain what women want?

She giggled. "No, but it does go some way to explaining the dynamics behind mate selection. As for what women want from a man, this woman wants chocolate and lots of it."

According to Surveys, These Are Women's Top Ten Erogenous Zones

1 Lips
2 Vagina/clitoris
3 Breasts/nipples
4 Wrists
5 Feet
6 Ears
7 Nape of the neck
8 Buttocks
9 Behind the knees
10 Inner thighs

"Men have to make love to feel loved. Women have to feel loved to make love."

(BILLY CONNOLLY)

"My favorite part of a man's body is the hipbone. I can die!"

(SALMA HAYEK)

"I think what makes a person sexy is probably their sense of humor."

(SIGOURNEY WEAVER)

Female painted turtles can store and use viable sperm for at least three years following mating.

- Female black widow spiders eat their husbands after mating.
- The female salamander inseminates herself. At mating time, the male deposits a conical mass of a jellylike substance containing the sperm. The female draws the jelly into herself, and in so doing, fertilizes her eggs.
- The female knot-tying weaverbird will refuse to mate with a male who has built a shoddy nest. If spurned, the male must take the nest apart and completely rebuild it in order to win the affections of the female.
- Male boars excite female boars by breathing on their faces.

Is It Different for Men?

If the "it" is cheating—which of course it is—then I refer you to the earlier answer. Yes, it is. It might even be necessary for the survival of the species.

There are biological imperatives and atavistic urges that do make it different for men, but does that justify such behavior?

My own view is that it doesn't. We may be programmed to behave in a certain way but that doesn't mean we shouldn't do our best to rise above it. I have a friend—more of an acquaintance really—who (at the last count) was boasting of three ongoing dalliances.

Even though I haven't identified him (though he'll know who he is), it might be fair to contrast him with another friend/acquaintance who *doesn't* cheat on his wife but loses no opportunity to run her down to anyone prepared to listen (aye, and many who aren't).

So, which is better—even if we can begin to define what we mean by the word "better"? The unfaithful husband who is loyal to his wife or the faithful husband who is disloyal to his wife? I certainly prefer the former—if only because I can't bear the latter's whining—but, ultimately, they're equally bad: that much infidelity amounts to disloyalty (it isn't so much the sex as the pillow talk and, worse still, the inevitable deceit), while that much disloyalty can only be corrosive to the marriage.

Perhaps we should devise a scale for infidelity. Lusting after women in your heart (*pace* Jimmy Carter) could be at one end of it, while *shtupping* the wife's sister and mother (as the British fascist Sir Oswald Mosley almost did—in fact, he screwed his first wife's sister and their stepmother, which is close enough) would be at the other end.

In fact, though (and while I'm in judgmental mode), I reckon that coming on to the wife's daughter from a previous marriage takes some beating in the caddish stakes. And yet, statistics show that when second or subsequent marriages—in which there are stepdaughters—fall apart, the biggest cause of the end of the marriages is the stepfather trying it on (one way or another) with the stepdaughter.

It certainly redeems Peter Sellers (who had sex with his first wife's best friend), Roald Dahl (who left Patricia Neal, his wife of thirty years, for her best friend, Felicity "Liccy" Crosland, whom he subsequently married), and the great womanizer Yves Montand—whom I met and found utterly charming (and who claimed, "I think a man can have two, maybe three love affairs while he is married. But three is the absolute maximum. After that you are cheating").

There is, however, another possible answer to the question. If the "it" refers not to cheating but to sex and emotions then, yes, it is different for men—as this wonderful gag that's been making the e-mail rounds illustrates perfectly:

Girl's Diary—Saturday, July 1, 2006

Saw John in the evening and he was acting really strangely. I went shopping in the afternoon with the girls and I did turn up a bit late so I thought it might be that. The bar was really crowded and loud so I suggested we go somewhere quieter to talk. He was still very subdued and distracted so I suggested we go somewhere nice to eat.

All through dinner he just didn't seem himself; he hardly laughed, and didn't seem to be paying any attention to me or to what I was saying. I just knew that something was wrong.

He dropped me back home and I wondered if he was going to come in; he hesitated, but followed. I asked him again if there was something the matter but he just half shook his head and turned the television on. After about ten minutes of silence, I said I was going upstairs to bed. I put my arms around him and told him that I loved him deeply. He just gave a sigh, and a sad sort of smile. He didn't follow me up, but later he did, and I was surprised when we made love. He still seemed distant and a bit cold, and I started to think that he was going to leave me, and that he had found someone else. I cried myself to sleep.

Boy's Diary—Saturday, July 1, 2006

England lost to Portugal on penalties. Got a shag though.

- In Uganda, it was once the case that if a woman's hand touched—even accidentally—any man's private parts (except her husband's), then that hand would be put in a pot of boiling oil. A second offense resulted in the "amputation of the offending hand."

"When women go wrong, men go right after them."

(MAE WEST)

- The Scottish poet Robert Burns had three daughters born out of wedlock, by three different women, and named each child Elizabeth.
- Turkey is the most adulterous nation in the world: 58 percent of Turks claim to have had an extramarital affair.
- When a school of baby catfish is threatened, their father opens his huge mouth and the youngsters swim inside to hide. When danger has passed, he reopens his mouth and lets the fry out. A father sea catfish keeps the eggs of his young in his mouth until they are ready to hatch. He will not eat until his young are born, which may take several weeks.
- On average, marriage increases a man's lifespan by 1.7 years, but decreases that of a woman by 1.4 years.
- The male fox will mate for life, and if the female dies, he remains single for the rest of his life. However, if the male dies, the female will hook up with a new mate.

Do Gays Talk Like Gays Because They're Gays or Do They Become Gays Because They Talk Like Gays?

You either accept the premise of this question—that *some* (please note) gay men talk in an effeminate way—or you don't.

One man who immediately knew what I was on about was Simon Fanshawe, a man who was one of the founders of Stonewall, the British organization that stands for "Equality and Justice For Lesbians, Gay Men and Bisexuals."

I've known Simon—on and off—for some sixteen years. Before we even met, I was a huge fan of his comedy. Nowadays, he's retired from comedy and has become a successful writer and broadcaster.

Simon is the same age as I and comes from a similar background: indeed, the only difference between us—apart from our sexual orientation—is that he's slimmer and I've got more hair. So, what did he make of my question?

"Some gay men talk like that. For them it's an unself-conscious distancing from accepted masculinity. Others talk like that because it brings with it a sense of identity. You see it with sixteen-year-old straight boys who are trying to be men and talk like (straight) men do.

"The history of being gay has been, for so long, the history of running away. With running away, you're trying to get away from the community you grew up in but you're also trying to find a new community and common ground. You're deracinating."

You're *what*?

"Deracinating. Look it up." (I did—"deracinate" means "To displace from one's native or accustomed environment.")

He continued. "You—we—were looking for men who were more feminine. Hence the voice. Hence the gay language. People converged on a way of speaking but then lots of cultures do that.

"Your modern gay loves all that camp because we dip in and out of it. It's a vernacular that's on my palate. It's something that men—gay and straight—can use.

"You should listen to yourself sometimes, Mitch, you can be a lot camper than me, you know."

Nonsense, I replied, but the more I thought of it, the more I realized he's right. My campest friends are all (as far as I know) straight.

And then there are the more modern gay stereotypes of extremely butch men dressed in leather—almost as a pastiche of the masculine stereotype.

Not to mention *Brokeback Mountain*.

No, Simon's right, there's been a lot of deracinating and it's affected both gay and straight culture.

But hang on, Simon, does the same go for mincing—you know the camp walk?

"Of course."

All right then, here's the $64,000 question: Are gays and lesbians born or made?

"Born. However, to believe there's a mono-causal reason is wrong: there isn't a single 'gay' gene, as the right-wing press would have us believe. You should read *A Separate Creation: How Biology Makes Us Gay* by Chandler Burr.

"However, even if we're born and not made, there's still no need to shock your parents. When you come out, do it discreetly. At Christmas, you say to your mother, 'Please pass the gravy to the homosexual.'"

> ◎ The word "homosexual" was invented by the Hungarian physician Karl-Maria Kertbeny in 1869.

What's the Origin of the "F" Word?

I'd always believed that it was an acronym—standing for "found under carnal knowledge"—that was written on the stocks into which adulterers were placed in Ireland.

Still, I thought I'd better check. So, with no great pleasure (as readers of *Why Girls Can't Throw* will understand), I turned to linguist and philologist Dr. Caron Landy.

She was as unlovely as ever—the single word "pardon," when I asked the question, speaking volumes for her low opinion of me—but once she got started, there was no stopping her.

"No, it doesn't stand for 'found under carnal knowledge.' Nor does it stand for 'for unlawful carnal consent,' nor for 'fornication under king's consent,' nor for 'forced unlawful carnal knowledge'—nor indeed for any other acronym you can come up with."

Aha, I said triumphantly, is this another example of a post-factum acronym?

"Isn't that what I just implied? In any event, the word you're searching for—and failing to find—is backronym. SOS is an example of a backronym, with people claiming it stands for 'save our ship' or 'save our souls'—when, in fact, it doesn't stand for anything."

That put me in my place.

"The origin of the word is surprisingly obscure. There are similar words in Middle Dutch, Danish, and Swedish dialect—meaning variously to thrust, copulate, push, strike . . ."

I got the drift.

"But it is probably safest to say that the word is mainly of Germanic origin. Here, for example, is what the *Oxford English Dictionary* has to say on the subject: 'Early modern English *fuck, fuk,* answering to a Middle English type *fuken* (weak verb) [which is] not found; ulterior etymology unknown. Synonymous German *ficken* can be shown to be related as well as Dutch *fokken*, meaning *to breed*.'"

Has it always been, you know, a shocking word?

"Yes. Its first known occurrence—in code, because it was so shocking—was in an English poem of the late fifteenth century, *Flen Flyys*, that sought to poke fun at the Carmelite friars of Cambridge. The context of the usage was, 'They (the friars) are not in heaven because they fuck wives of the town of Ely.'"

Fascinating.

"In fact, the word has been considered so shocking that the *OED* wouldn't include it in the dictionary until 1972."

Ever since when naughty young boys have been looking it up and giggling.

"If you say so."

☺ The word "sex" was coined in 1382.

Why Did People Think That Masturbation Made You Blind?

Before I (attempt to) answer this question, let me just say that I entirely agree with Woody Allen, who described the old five-finger shuffle as "sex with someone you love." And who could disagree with Truman Capote (a tiny little tosser), who pointed out that "the good thing about masturbation is that you don't have to dress up for it"?

In fact—and this is relevant to the question if not to the answer—I often joke that although I came to it (so to speak) late, I wanked so much between the ages of sixteen and eighteen that I very nearly lost my eyesight. I always say this totally seriously and—for a split second, at least—people seem to believe me.

Well, obviously you can't get blind from too much wanking. Can you? Don't be stupid, Mitch, of course you can't! Are you sure? Yes. A hundred percent sure? Yes. Stake-the-mortgage-on-it-hundred-percent-sure? All right, let me check.

So, once again (see *Why Girls Can't Throw*), I turned to Dr. Roland Powell, an MD of my acquaintance, for reassurance.

"Are you kidding me?"

Well, no, not really.

"All right then, let me put your mind at rest: there is not the slightest chance of you—or anyone else—losing their eyesight as a result of excessive masturbation."

What, even if one were wanking while crossing the road?

He ignored my idiotic attempt at witticism. "In fact, the only possible adverse consequence of excessive masturbation would be chafed and sore genitalia."

As we were talking, I found myself singing The Vapors' song "Turning Japanese," which is reckoned to be about the joys of self-abuse because men screw their eyes up—thereby making them look Japanese—as they reach the vinegar strokes (or "climax"). Perhaps, I suggested to the good doctor, this explains the fears about blindness?

"I'm not sure I follow your train of logic but it's clear that those fears were stoked by misguided people who were desperate to stop boys from masturbating."

I did some further research into this and came across a doctor named Samuel August Tissot, who, in 1760, wrote "L'Onanisme, ou Dissertation Physique sur les Maladies Produites par la Masturbation" (you can translate it yourself). This cheerful little tome—which was still being published in the twentieth century—aimed to scare generations of adolescent boys into sleeping with their hands above the sheets. Here's Tissot writing about a home visit he made to a poor fellow supposedly suffering from "post-masturbation disease":

"I went to his home; what I found was less a living being than a cadaver lying on straw, thin, pale, exuding a loathsome stench, almost incapable of movement. A pale and watery blood often dripped from his nose, he drooled continually; subject to attacks of diarrhoea, he defecated in his bed without noticing it; there was constant flow of semen; his eyes, sticky, blurry, dull, had lost all power of movement; his pulse was extremely weak and racing; labored respiration, extreme emaciation, except for the feet, which were showing signs of edema. Mental disorder was equally evident; without ideas, without memory, incapable of linking two sentences, without reflection, without fear of his fate, lacking all feeling except that of pain, which returned at least every three days with each new attack. Thus sunk below the level of the beast, a spectacle of unimaginable horror, it was difficult to believe that he had once belonged to the human race. . . . He died after several weeks, in June 1757, his entire body covered in edemas."

And all that just from too much time spent playing the pink oboe.

As a result of scaremongering from Tissot and others like him, there was plenty of interest in "cures" to stop youths from abusing themselves. Dr. Benjamin Rush, a signatory to the U.S. Declaration of Independence, recommended "a vegetable diet, temperance, bodily labor, cold baths, avoidance of obscenity, music, a close study of mathematics, military glory, and, if all else failed, castor oil."

John Harvey Kellogg, the cereal manufacturer, was dead set against masturbation, which (or so he believed) caused idleness, abnormal sexual passions, gluttony, and sedentary employment.

Sounds about right.

Kellogg—sorry, Dr. Kellogg—said of masturbation, "The most loathsome reptile, rolling in the slush and slime of its stagnant pool, would not so demean itself," and believed that a bowl of cornflakes would help to take a young man's mind off it.

If that didn't work, Kellogg favored the use of an electrical urethral probe in fifteen-minute doses, two or three times a week. He also made use of the "bougie," a long, flexible tube made of rubber or metal that was inserted into the male urethra, often delivering a caustic dose of burnt alum or silver nitrate or some other noxious substance.

Kellogg also recommended bandaging boys' genitals, and/or tying their hands to the bedposts at night.

Other contemporary deterrents included the application of leeches to the inner thighs, confinement in straitjackets, wrapping in cold, wet sheets while sleeping, burning genital tissue with a hot iron, metal gloves, infibulation, urethral rings of metal spikes that would stab the penis if it became erect, and various anti-masturbation contraptions—like the genital cage that used springs to hold a boy's penis and scrotum in place and the device that sounded an alarm if a boy had an erection. If all else failed, there was always castration.

Girls weren't excluded. There were metal vulva guards and, just to be safe, clitoridectomies (a disgusting practice still carried out in parts of Africa).

By comparison with those, other supposed cures—like potassium bromide, hot pepper, hydrochloric acid, or mercury—seem relatively civilized, but perhaps not as much fun as the opium or marijuana that some boys were given to take their minds off masturbation.

Mind you, it was difficult for them to avoid thinking about it—especially as Victorian parents would take their children to wax museums just to see the gruesome "effects" of post-masturbation disease.

Meanwhile, schools were equipped with anti-masturbation benches designed so that a young man's thighs were not brushed together in a way that might excite him. Who knows? Such a bench might have stopped André Gide, the French author, who was expelled from school for wanking in class.

This obsession with—against—masturbation can perhaps be explained by the Victorians' belief that we only had a limited amount of semen, which shouldn't be wasted through masturbation but should instead be saved for our wives to make the sons and daughters who would run the British Empire. These

Victorians believed that this reservoir of cum was located at the base of the spine. Obviously, they were absolutely mad, but you can see what led them to believe that that was where it would be, as every time a man ejaculates, he feels a spasm in just that place. Well, this man does, anyway.

In fact, the conviction that there was a set number of ejaculations for each man persisted well into the twentieth century. Ernest Hemingway, for example, earnestly (*sic*) declared, "The number of available orgasms is fixed at birth and can be expended. A young man should make love very seldom or he will have nothing left for middle age."

Er, no.

Looking back, it seems extraordinary, the lengths to which otherwise intelligent men went just to stop boys doing what comes naturally, but I guess it has to be seen in the context of a repressed age in which *any* form of self-indulgence or self-expression was frowned upon—to the extent that *any* abuse justified the prevention of self-abuse.

◎ 95 percent of men masturbate, compared to 70 percent of women.

Prodigious Masturbators

Howard Stern
Hans Christian Andersen
André Gide *(once expelled from school for wanking in class)*
Yukio Mishima
Friedrich Nietzsche
Vaslav Nijinsky
Bill Tilden
Jean Genet

Is Bestiality Immoral?

Absentmindedly, I Googled the three words of the question and got just four results, but one of them was very interesting: in a blog named the Daily Kos, a man named Neil Sinhababu posed just that question. I contacted him and obtained his permission to share the results with you.

He wrote: "I've never thought there was anything immoral about a human being having sex with a nonhuman animal. Not that I'd ever do that sort of thing, but if I found that someone I knew had done it, my feelings would be more of surprised amusement than of moral disapprobation. I'm curious about what other people think about this."

He went on to qualify this by writing: "I'm talking about cases of bestiality where there isn't any harm to either the animal or the human being, and both parties have a good time. Let's also consider cases where no new diseases make the jump across species as a result of the sexual encounter.

"As a utilitarian about morality, I think that actions are good in virtue of their promoting the general happiness and decreasing suffering, and bad in virtue of their decreasing happiness or increasing the amount of suffering in the world. (Complications arise when people intend good things but bad things result, and vice versa. But this can be kept aside from the bestiality issue.) Utilitarianism is compatible with my view on the bestiality case, since I think the possibility of pleasure and the absence of harm mean that sex with animals is okay. However, I imagine that most people disapprove of bestiality in a way that utilitarianism can't account for."

After reading such a mature, intelligent, philosophical argument, you'll be as surprised as I (and, no doubt, Neil too) was that out of the 106 people who voted, a massive 86 percent (i.e., 92 people) thought it was immoral, 6 percent were unsure, and only 6 percent—seven people out of 106—thought there wasn't "something morally wrong about a person having sex with an animal."

For my own part, I admire Neil's utilitarian argument but I would vote with the majority, because, as Neil concedes, utilitarianism doesn't cover this issue. Indeed, it was Jeremy Bentham himself who, realizing that utilitarianism couldn't be applied to animals, said instead, "The point is not how clever they are but do they feel pain, can they suffer?"

On the grounds of suffering, there's no evidence to suggest that bestiality is any worse for an animal than, say, intensive farming followed by a trip to the abattoir; indeed, it is almost certainly preferable.

Which leaves us with the human participants. As you know, my general approach to matters of sex is: whatever consenting adults consent to do in private. With bestiality, one has to ask how informed and reasonable the consent is. Now add to the mix the undeniable fact that bestial images are widely used in pornography with the clear intention of demeaning the human (invariably, a woman), and I would say that the whole issue becomes tainted to such an extent that it renders bestiality itself—even in the abstract—immoral.

There is one qualification to this. If bestiality is sex with animals, zoophilia is (or so its practitioners would have it) love—albeit sexual love—with animals. I understand the distinction but would still maintain that interspecies sex is "a bad thing."

Fortunately, human-animal couplings are rare. However, I can tell you that the porn star Linda Lovelace had sex with a dog for a porn film, that Algernon Swinburne used to do it with his pet monkey* (which he later killed and ate when it bit one of his guests), and that Larry Flynt had a sexual experience (in 1948) with a chicken, which he wrote about in his book *Sex, Lies and Politics: The Naked Truth*.

◎ In Peru, there is an old law which prohibits single men from keeping a female alpaca in their homes.

* This may not be true, as the rumor was said to have been started by Swinburne himself.

Can Open Marriages Work?

Like most people, I've a tendency to universalize my own experience—especially when that experience is germane. So, let me start my answer by saying that I've been happily married for twenty-three years, but that a big reason for my—our (I hope)—marital happiness is that it's a monogamous relationship. Before meeting Penny (my wife), the longest I ever lived with a woman was for forty-eight hours—and that was only because I put my back out. Pre-Penny, I was, to put it mildly, a bit of a slut; since then, I've been a one-woman man (in deed if not always in thought). It's not a question of fidelity (a much overrated virtue: give me loyalty or, better still, kindness any day) for the sake of fidelity, but because, for me, fooling around would be the thin end of the wedge. As the late Richard Burton said, "The minute you start fiddling around outside the idea of monogamy, nothing satisfies anymore." I'm pretty sure that's how it would be for me.

And yet it would be foolish to pretend that there's only one path to happiness—especially as, according to anthropologists, the survival of the species is predicated on men scattering their seed far and wide. After all, as Somerset Maugham pointed out, "You know, of course, that the Tasmanians, who never committed adultery, are now extinct."

Some men even go so far as to claim that the desire to sleep with as many women as possible is hardwired into the male psyche: I can remember George Burns saying, "If you were married to Marilyn Monroe—you'd cheat with some ugly girl."

As we've already established, a woman has a different anthropological imperative: she needs to find a man who will nurture and provide for her and her baby. Interestingly, this might not be the same man who made her pregnant in the first place. The evidence suggests that woman seeks the biggest alpha male (aka the biggest bastard) to impregnate her and then the most suitable man to look after her and the baby. Inevitably, in modern society, she tries to find a man who combines both qualities, but her primeval need to keep her man runs deep.

So, with women programmed to cheat before becoming mothers and men programmed to cheat after becoming fathers, it's a wonder the divorce rate isn't even greater than one in three (like three in three). The fact that it isn't is a tribute to love, trust, and the human ability to shake off atavistic conditioning.

In this context, an open marriage could seem less like a total indulgence and more like a sophisticated arrangement to safeguard the structure—albeit not the substance—of marriage.

My friend Tim and his wife Jill have an open marriage. Although Tim and Jill miss very few opportunities to bang on about their—yawn—love lives, I have changed their names for the sake of their children (one of whom is just twelve years old). They've been "open" for over two years and they say that far from damaging their relationship, it's actually enhanced it. They even go to swinging parties together and swap partners. Personally, I can't think of anything more ghastly, but they seem to thrive on it—or, at least, Jill does. Because just as it's rare to find two people in a marriage with precisely the same sex drive, so too is it unlikely that both partners will have the same desire to open up their relationship to outsiders. Consequently, in every open marriage, there has to be an instigator. Ordinarily, for obvious reasons, it's the man, but, in this instance, it was Jill's idea. After discovering that Tim had had a one-night stand while working away from home, she decided to get even by shagging the architect who was doing their extension. Apparently, she enjoyed it so much that she told Tim that that way lay their marital salvation. It seems to work for them, though I have my doubts. For starters, they're far too evangelical about it—you know, "come on in, the water's lovely!" (and no, I don't think they're trying to recruit us personally)—and, secondly, the time will surely come when either (a) one of them—most probably Tim—will decide that, like Mrs. Patrick Campbell, they yearn for "the deep, deep peace of the double bed after the hurly burly of the chaise longue" (you'll note that I have a somewhat romanticized view of adultery) or (b) one of them—most probably Jill—will find someone who represents more than just a quick lay and they split up the marriage to go off with that person.

Who knows? Meanwhile, they're both getting plenty of great sex. So, who's the idiot?

Couples Who Had Open Marriages

Carl and Emma Jung
John F. and Jacqueline Kennedy
Leonard and Virginia Woolf
William and Jane Morris
Harold Nicolson and Vita Sackville-West
Dmitri and Nina Shostakovich

Gary and Veronica (Rocky) Cooper
Sir William and Lady Emma Hamilton
Horatio and Fanny Nelson
Marlene Dietrich and Rudolf Sieber

What conclusions can we draw from this list (other than that I'm a damn fine researcher—and more than a little strange)? Some of the marriages survived, some thrived, some died. Perhaps, opening up the marriage postponed the death rites in that last category. But what do you notice about nearly all of these names? Yup, they're all dead. This isn't entirely due to my healthy respect for the laws of libel. It's because I suspect that the whole question of open marriages harks back to the days when marriage itself was de rigueur and divorce was too stigmatic and expensive to contemplate. Now that people can and do "live in sin" and have multiple sequential partners without attracting very much adverse comment, who needs an open marriage?

"I am a one-woman guy and I am blissfully happy with Iman. She's the only girl for me."

(DAVID BOWIE)

◎ Beavers mate for life.

"Like everything which is not the involuntary result of fleeting emotion but the creation of time and will, any marriage, happy or unhappy, is infinitely more interesting than any romance, however passionate."

(W. H. AUDEN)

- Couples who marry in January, February, or March have the highest divorce rate.
- Couples are more likely to break up on January 12 than on any other day of the year.

If I Decide to Become a Slave to a Dominatrix, Who Pays Whom?

The clue to the answer is in the question: if I decide . . .

Obviously, if it's something I want to do, I'm going to have to pay for it. But why would I pay to be someone's slave? Surely the whole point of slavery is that one is owned—not that one pays to be owned? Doesn't that upset the delicate balance of the master-slave relationship? I mean, how would the slave trade of the eighteenth and nineteenth centuries have operated if slaves had to pay their masters? Simon Legree would have been out of work (no bad thing).

I was fortunate enough to strike gold with the first dominatrix I found: Mistress Mia, who had a listing that reads:

> *I Mistress Mia am looking for a special little slave, you will look after my every need in the house, you will take me shopping and whatever i demand of you. if you can not stay fri to sun then dont bother, if you can what are you waiting for slave mistress wonts to here from you. call me on . . .*

So, I called her and told her about this book and asked the question.

She laughed. "They pay me!"

How much?

"A hundred and twenty pounds an hour."

Not for the whole weekend, surely?

"No, obviously it's less for the whole weekend."

But even so, that's an awful lot of money just to do your shopping and carry out your demands. Sounds to me like any married man's regular Saturday morning.

She giggled.

I was about to ask the $64,000 question—or maybe that should be the £120 question—why do they do it? Then suddenly I realized that I knew why they did it. It's about control issues and working through fantasies, isn't it?

Mistress Mia confirmed that it was. "There are a lot of men out there who . . ." Her voice trailed off but I caught her drift.

For some reason, I felt ashamed. No, it wasn't "some reason." Here I was again, making fun of other people's needs/problems/outlets. Just because I have no personal need for Mistress Mia doesn't mean I should deride those who do.

We talked for quite a lot longer. She's a fine woman (mid-forties, I should say), who is tolerant (and sometimes fond) of her clients. "I'm a dominant woman by nature. I'm very assertive and men just love that. And the older I get the more assertive I get."

Good point: being a dominatrix means never having to say you're retiring. The older a dom gets, presumably the more appealing she'll be—at least to some blokes.

I looked her up again on the Internet.

hello slaves and all u subs, on ur knees before ur goddess, good, now in session in dover wed and sat 11 to 11. i have use of a fully equipped dungeon. i cater for most fetishes, cp and bodywor-

ship. so come crawl to a goddess and let me kiss ur lips and enter ur soul

OK, so her grammar and syntax aren't up to much—but she certainly gets her message across.

And while I might not care for being corporally punished in a dungeon (however well equipped)—indeed, I would pay good money NOT to be so treated—it's evident that there are several men who do. You might even be one of them.

Good luck to them and, if appropriate, to you too.

Men Who Enjoyed Being Dominated

T. E. Lawrence (*liked to be flogged*)
James Joyce (*loved being dominated—and perhaps whipped—by Nora Barnacle, his lifelong companion*)
Jean-Jacques Rousseau (*a masochist who liked to be spanked*)

Consider also **Sylvester Stallone**, who had sex with a girl extra on the set of *Judge Dredd* while he was still wearing his microphone, and was heard saying "slap my butt." The next day, the crew wore T-shirts saying "slap my butt."

We Know That Men Become Women but Do Any Women Become Men?

I am privileged to number among my friends a fine young man named Bryn . . . who used to be a fine young woman named Bee.

Anyway, Bryn tells me that there are probably as many transsexual men (that's women who become men) as there are transsexual women (that's men who become women). However, when it comes to medical treatment—including drugs and operations—transsexual women outnumber transsexual men. On the other hand, in Poland, the majority of sex-change operations are female-to-male rather than male-to-female.

According to Bryn, "The main reason why relatively few transsexual men have medical help to live as men is because, in our society, it's much easier for a woman to live the life of a man than it is for a man to lead the life of a woman."

While I pride myself on my tolerance, I confess that I'm not—how can I put it?—the most fashionable of men, but I've had no difficulty whatsoever accepting Bee as Bryn.

I think it's because although "she" wasn't especially butch, her squat, shapeless body, combined with a severely cropped hairstyle, made her look an improbable woman. Having said that, she always had a beautiful voice which, fortunately for him, is still very fine (albeit in a lower register). When she told me that she was becoming a man, I wasn't really very surprised. In truth, I felt relieved.

You see, Bee was a lesbian, and although I had absolutely no problem with that (my stepsister is also a lesbian), I always felt I had to watch what I said.

This wasn't because Bee was at all touchy—au contraire—but because, even at the best of times, I can come across as a potty-mouthed oaf. Consequently, not only was I careful never to refer to lesbians as "lezzers" or "dykes" (words that, in any event, rarely pass my lips), but I also had to modify my—ahem—somewhat misogynistic views.

Conversations with Bee could never start with lines like "Women, huh!" or "Bloody women, can't live with them, can't shoot them."

In short, there was an element of treading on eggshells.

Then Bee gave me the news . . . and suddenly, I could say what I liked!

And you can be sure I did.

Bryn, however, is much more measured and civilized than I am. Although, as a convert (as it were), he might be expected to embrace his new gender so wholeheartedly that he ends up as a stereotypical "lad" or he-man, he is, in truth, no different than he was as a woman: a thoroughly decent human being. Nevertheless, journalistic integrity compels me to report that he did recently point out to me that one should never trust any creature that bleeds for a week every month without dying.

The differences between Bee and Bryn are much more subtle than one might have expected: he looks much more impressive as a man than he did as a woman, and although he has maintained the same weight (less the weight of his breasts), he looks much slimmer and fitter. There's also the sense—or, at least, this is what I've inferred—that he's much happier in his own skin.

In short, Bryn is a top bloke.

- There has never been a sex-change operation performed in Ireland.
- It is estimated that a thousand people seek sexual-reassignment surgery each year and as many as forty thousand postoperative transsexuals live in the USA. Experts calculate that there are at least three to five times as many transsexuals who don't have surgery.
- 40 percent of transsexual people who don't have sexual-reassignment surgery are either institutionalized or die prematurely. Fewer than 1 percent of those who do have surgery subsequently regret it.
- When a shrimp is first born, it is male, but it gradually evolves to female as it matures.
- The clown fish has the ability to change its sex. If a breeding female dies, the male fish will change its sex and mate with another male. A sea polyp can also change sexes at will.

. . . and Following on from That: When a Lesbian Has a Sex-Change Operation to Become a Man, Does Her Erstwhile Lesbian Partner Cease to Be a Lesbian?

Once again, I'm fortunate enough to have Bryn as an expert guide.

As Bee, he lived in a lesbian relationship with Sarah.

Now that Sarah is living with a man in a sexual relationship, has she become, ipso facto, heterosexual?

"No," says Bryn, clearly enjoying the intellectual challenge of the conundrum (see, he really is a man!). "She's still a lesbian. The fact that she was in love with Bee and is now in love with Bryn doesn't reflect a change on her part but on mine. I'm the one who's changed, not her."

But still . . .

"Yes, I know what you're saying. Put it this way, she loves me—well, hopefully anyway. That's me, Bee, Bryn, or whatever label you want to put on me, but I'm essentially the same person I always was—whatever my gender."

That's fair enough.

"And the point is that if we ever split up . . ."

. . . Perish the thought . . .

". . . Absolutely, Sarah would still be a lesbian and would, no doubt, get together with another woman."

I understand, I told him.

And, incredible as it might sound, I really do.

◎ Most snails are hermaphrodites, meaning they have both female and male reproductive organs.

Has Any Man Ever Gone to a Prostitute and Found Someone He Knew?

That's got to be a nightmare, no? Fellow goes to a prostitute and finds his wife/sister/wife's sister—well, you fill in your own unbelievably horrid scenario. One thing's for certain, it would lead to instant detumescence. No matter how much Viagra you might have swallowed.

I asked around—checked with the usual suspects—but drew a complete blank. In fact, none of my crowd even admitted to having ever been with a prostitute—let alone found one he already knew.

This is manifestly absurd. There are so many tarts in Britain that, by the laws of probability, some of my mates must have flashed the cash. And don't let the denials that "I'm a happily married man" fool you: I remember being accosted by a pimp in Soho a few years ago. As he tried to interest me in the photos in his folder (those were the days before mobile phones/cameras), I protested that I was—yes—"a happily married man."

"Happily married?" he said, his face breaking into a big

grin. "If it wasn't (he didn't, as you and I would have done, employ the subjunctive) for married men, I wouldn't have any clients."

So there.

Nor is prostitution uniquely rife in Britain. In the U.S., it is estimated that twenty-three people in every hundred thousand are prostitutes (or 0.023 percent of the population). Not an insignificant number.

Anyway, having failed so miserably with my friends, I logged on to the fabulously comprehensive (sui generis) site of Punternet.com. What an eye-opener! You should go there—if only to read the "field reports" which are (evidently) so well researched and, for the most part, well written that they would put consumer journalists to shame.

Perhaps the most extraordinary thing the site reveals is the prevalence of prostitution in Britain.

Alas, even after reading hundreds of reports, I hadn't found anyone who'd got the shock of their lives (I guess they wouldn't exactly rush to write about it if they had).

So, I logged on to some of the (very) many chat rooms devoted to prostitution—home and abroad—and ran into a very different problem: the assumption that I was a wanker.

All right, so I am, but not in that sense. The questions I was asking were genuine—as you must surely concur. Trouble was, no one—neither the girls nor their clients—believed me. And the more I protested my innocence, that I was a professional writer, the more they told me to "go away, tosser."

But I'm Mitchell Symons, the author of *Why Girls Can't Throw*, just go to Amazon.com and look me up!

"How do we know you're him? You're probably making it up."

But obviously I'm him—me—why on earth would I pretend to be him—me? It defies belief!

Therefore, in the interests of research—honest!—I decided to pay a visit to a brothel and talk to a professional, face-to-face. It's what we in the writing biz call "fact-finding" and is perfectly valid—even if I did have the devil's own job explaining it to my wife, who, not unreasonably, asked me why I was insisting on a primary source for this question when I was perfectly happy using secondary and tertiary sources for all the others ("nicking it off the Internet" were her precise words, as I recall).

Having looked up a likely brothel (in all but name) in the local paper, I drove to a rundown part of Brighton, parked, and found myself outside a grotty house.

The middle-aged woman who answered the door was wearing a badly fitting tracksuit and leggings, and, with a voice that betrayed a lifetime's devotion to cigarettes, bade me "come inside." Blimey, I thought, if she's one of 'em, I'd rather go with Lorena Bobbitt. What's that line in *Saving Private Ryan*? "She fell off an ugly tree and hit every branch on the way down?" Yup, that was her.

But it turned out that she was only the maid and, I later concluded, probably a former prostitute herself. Not what you'd call a walking testament to the benefits of a life on the game.

I looked around the dingy room into which she'd ushered me, asking me to wait while she rounded up all the spare girls and sent them in for my perusal. The room wasn't what you'd call over-furnished: queen-sized bed, bedside table with a box of tissues and a broken lamp-stand, wooden chair, and a chest of drawers with a stack of top-shelf magazines on it—all well-thumbed through, by the look of them. In the corner was an open wardrobe where I could see a number of uniforms from the stock list of (supposed) male fantasies—nurse, schoolgirl, policewoman.

A girl wearing a negligee knocked and entered the room. Well, I say a girl, she was thirty-five if she was a day: overweight, bad skin, and not much prettier than the maid. She'd be lucky to give it away, I thought while simultaneously chiding myself for my lack of charity.

"My name's Candy," she said, in a surprisingly posh accent, and left the room.

The next girl was again overweight but she had a decent cleavage. "I'm Layla," she said, offering me her hand. It somehow seemed an incongruous greeting in what was, after all, a whorehouse.

The third girl—Zoe—was young (twenty-three, as I later discovered), attractive, and looked extremely fetching in a black basque. I decided to pick her—not, if I was being honest, because she would have a wealth of experience to draw upon, but because I didn't think it was inconsistent with either my mission or, indeed, my marriage vows (not that I took any, as we were married in a registry office), to talk with an attractive girl for a few minutes.

There was no one else available—the other three of the six girls I had been promised on the phone when I had called for the address were presumably busy—and so I told the maid that I'd like to see Zoe.

"So what would you like?" asked Zoe.

"Can we just talk?"

"Sure," she replied, although I could see her thinking *We've got a right one here.*

"No, seriously. I'm writing a book—it's the follow-up to my last book, *Why Girls Can't Throw . . .*"

"I can."

"What?"

"Throw! I can throw."

"Ah yes, quite possibly. In fact, I answer that question in

the book and it turns out that girls can indeed throw when they've been taught." I suddenly remembered where I was. "So what would you charge for your time?"

"Well, if you really want to talk, I'll charge you forty pounds for thirty minutes. Is that OK?"

I handed over the money, and she disappeared only to return a couple of minutes later.

"Where do you want me?"

"I don't mind," I spluttered. "Perhaps if I sit in the chair and you sit on the bed?"

"All right."

And, blow me down, she started to peel off her basque. "Excuse me!"

"Yes?" she said, with her really rather excellent breasts on display.

I was torn between not wanting to appear censorious and not wanting to be an old lech. Well, which do you think won? And so, I asked her whether she'd ever met anyone she already knew in her capacity as a . . .

"Working girl."

"Yes, as a working girl."

It turned out that she hadn't—partly because she wasn't from Brighton—but that her friend Emma, a local girl, had entertained a man whom she subsequently met at a family gathering.

"And was it embarrassing?"

"No, not for her. She wasn't the one who was married. But he nearly had a nervous breakdown on the spot!"

"How would you feel if it happened to you?"

She tilted her head to one side and gave it some thought. "I dunno. I mean, I'm married but my husband knows what I do."

I was intrigued to discover she was married. "How come he doesn't mind?"

"We need the money and he doesn't earn very much."

"But doesn't it, er, have an adverse effect on your relationship?"

"Nah."

There wasn't much I could reply to that, and so I asked her more about herself and her work, taking notes in my (extremely poor) shorthand as she spoke—only occasionally looking up to catch a glimpse of her . . . well, never mind what, let's just say that sometimes there are some perks to my work.

"Trick is to make the punter relax—be pleasant to him—and he'll be nice back. Sometimes it's over in seconds and then you've earned forty quid for a minute's work. Where else could you get that much? And it's tax-free. At my last place, I made three hundred and forty quid in one night. Now that's good money.

"I don't do anal," she said apropos of nothing. "Some of the girls do but I don't. I also don't do French [oral] without a condom. That's not because I'm squeamish or anything, it's because I'm frightened of getting chlamydia."

"Can you get it like that?"

"Yes. You get it from unprotected sex. It doesn't do anything to you at the time but it can make you infertile later on. I've told one or two of the other girls at the other place I work about it and they've stopped doing any unprotected sex—except maybe breast relief.

"I go to a clinic regularly, where they know I'm a working girl and they give me priority. It isn't just chlamydia, there's also genital warts. They're contagious too. You can't be too careful."

Hmm . . . not quite the conversation I'd imagined. "Tell me," I said, "do you derive any pleasure from your work?"

She smiled—possibly at my attempt at lofty detachment.

"Do you mean have I ever come? Yeah, once or twice."

We exchanged a few more pleasantries and then the time was up and we were shaking hands. "Come back and ask for me," she said, leaving me in no doubt what she meant.

I pretended to misunderstand her. "It's OK, I've got everything I want."

So, that's the question answered—sort of—and a few others besides. One thing my (albeit) limited experience did give me was a respect for prostitutes—or at least *that* prostitute (if not for prostitution): an understanding that they're human too and not just toys for boys.

One final thought on the question: there are plenty of men who have gone to prostitutes, who have discovered that they (the pros) knew them (the punters). Those are, of course, the ones who find themselves on the front pages of the tabloid press. But they're not alone.

The following men all had sex with prostitutes—some of them have even boasted about it: **Elvis Presley, Charlie Chaplin, Duke Ellington, Vincent van Gogh, Clark Gable, Jean-Paul Sartre, Errol Flynn, Victor Hugo, Henri de Toulouse-Lautrec, Oscar Wilde (contracted syphilis from one when he was at university), Casanova, Honoré de Balzac, Johannes Brahms, Lord Byron, Fyodor Dostoevsky, King Edward VII, F. Scott Fitzgerald, George Gershwin, Ernest Hemingway, Friedrich Nietzsche, Grigori Rasputin, Hans Christian Andersen** (who used to go to brothels to *talk* with the girls), **Dudley Moore, Al Capone, and Keith Moon**.

In addition, the following men all lost their virginity to prostitutes: **Henry Fonda, David Niven, Anton Chekhov, the Duke of Windsor, Napoleon Bonaparte, James Boswell, James Bond** (in Paris, at age sixteen, with a prostitute named Martha Debrant), **James Joyce, John F. Kennedy** (the girl charged $3), **Benito Mussolini, Leo Tolstoy, H. G. Wells,** and **Groucho Marx.**

I couldn't find any women who'd lost their virginity to male prostitutes, but the following men all paid for sex with male prostitutes or rent-boys, and some of them were "outed" by men they'd paid: **George Cukor, Howard Hughes, Roy Cohn, A. E. Housman, Charles Laughton, Joe Orton, Leonardo da Vinci, Monty Woolley, Cole Porter, Pier Paolo Pasolini, T. E. Lawrence, Cecil Beaton, Errol Flynn, W. Somerset Maugham, Montgomery Clift, André Gide, Christopher Isherwood, Yukio Mishima, Oscar Wilde, Lord Alfred Douglas,** and **Brian Epstein.**

- In Babylonia, in the tenth century B.C., prostitution was compulsory.
- Why is prostitution legal all over Nevada but not in the cities of Las Vegas and Reno? Because research shows that men gamble more when they take their wives with them.

. . . and Why Do Prostitutes Offer French but Not Danish?

French is, as we know, oral sex, while Danish is short for Danish pastries.

I didn't ask Zoe whether she offered Danish, but having seen the shabby surroundings of her workplace, I very much doubt it.

There are a lot of silly questions in this book that manage to provoke really rather interesting answers.

I don't think this is one of them.

Obviously, there is a certain amount of fun to be had with euphemisms, but it's all a bit old-hat. The days when prostitutes would advertise "French lessons" or put up cards saying "Model seeks unusual positions" are long gone. Nowadays, it's much more in your face—whether it's on the Internet or on a card in a telephone box.

Maybe directness is a good thing—though not, I'd have thought, if kids could trip over it.

Taking the question at face value, I was surprised to see just how much some of the brothels (in all but name) offer on their Web sites. Jacuzzi baths, drinks, showers—who knows? maybe you can get pastries as well.

The thing to remember is that prostitution is capitalism at its rawest and most unfettered: whatever a person's prepared to do in return for what another person's prepared to pay.

It may be ugly—and, at the bottom end of the market, where girls are lured from Eastern Europe and then effectively kept as sex slaves, it undoubtedly is. However, whatever free adult citizens choose to get up to in private isn't anyone else's business but theirs.

I asked one punter I found online why he paid for sex and he told me that he had girlfriends but that he preferred going with tarts. "Apart from anything, it's cheaper," he said, echoing the old saying that "the big difference between sex for money and sex for free is that sex for money usually costs less." In other words, with prostitutes, men are paying for the sex and the privilege of not having to undergo postcoital intimacies. Or, as they say in Hollywood, you don't pay a whore to come, you pay her to go.

"I believe that sex is the most beautiful, natural, and wholesome thing that money can buy."

(STEVE MARTIN)

Does the Etiquette of Swinging Extend to Asking Your Buddy's Wife If She'll Iron Your Shirt?

I've already shared with you my opinions on "open marriages," so you won't be too surprised to learn that Mrs. S. and I don't go in for wife-swapping or, as it's sometimes known, swinging. Besides, it wouldn't be fair: the man who got my wife would be a hell of a lot luckier than his missus would be (notwithstanding my fascinating personality, I am more mattress than man).

Which was why I was extremely lucky to find Dave and Linda, a couple who organize, er, get-togethers at their house in East Sussex (Skinnydippersuk.com/index2.html).

The first thing Dave corrects me on is my terminology. "We don't like the word 'swinging,' we prefer 'adult parties.' But our parties are not just about sex. If you go to a party where you don't fancy people, it's boring. We come from the music scene, so we just took that scene into this scene. There's no pressure.

"We do it because we believe in it. It's a private house, which cuts through the club stuff. We're not technically restrained by anything. We like to party. There's not much money in it but obviously you have to stay in business.

"I grew up in the sixties. I try to create an atmosphere of freedom. The golden rule is: so long as a couple goes home together. Don't come here if you're in trouble. You've got more chance of losing your partner at Tesco's—"

Not at our local branch, I mutter.

". . . than you have at one of our parties. If someone does lose their partner at one of our parties, they would have done so anyway."

That sounds as though you're denying any cause and effect between swinging—sorry, "partying"—and splitting up.

"Coming into this scene can show up the flaws in a relationship and it accelerates fissures caused by those flaws. People who come along have flaws and this might be a way of working through."

Hmm.

And then, blow me down—stop tittering there, in the back—Dave and Linda invite me to come along.

What, on my own?

"On your own or with your wife."

Gulp.

Have a look at some of the people on the Web site. I did and they all—or mostly all—looked perfectly normal. Whatever normal means.

"Look, we get between a hundred and a hundred and eighty people to our parties. Mostly couples but also singles. Friday nights, we have singles as well as couples."

Mostly men, these singles?

"No, there isn't a surplus of men."

Another cherished myth exploded.

It's tempting—it was—but it's not really me.

"Sure?"

NO! But I'm sticking with that. Besides, Dave and Linda, with all this super information you're giving me, you're neatly managing to evade the question: I know that at one of your parties, a man can proposition a woman—and vice versa—but what if he were to ask her to iron his shirt?

Understandably, Dave and Linda were somewhat less forthcoming on this (hitherto unexplored) aspect of adult partying, and I was starting to feel like every sort of silly ass when an extraordinary thought occurred to me: ask an impertinent question and you're halfway there to finding a pertinent answer.

Of course I had meant the question to be silly and facetious but, in fact, there's an important truth in there: if you reduce sexual intercourse, the ultimate act of intimacy, to the level of a parlor game to be played with multiple partners, then you need to find something else to replace it. And, who knows, performing minor chores for each other might be all that you've got left.

Having said that, to each their own, and East Sussex is tantalizingly close to where I live . . .

Why Do Men Fall Asleep after Sex?

Well, I do. Don't all men? Why? Presumably because we have sex last thing at night. Also, as men, we're task-oriented: job done, let's move on to the next one, which, after a good fart, is going to sleep. The cuddle can wait for another day.

Having said that, it is undoubtedly true that sex—accompanied or solo—does make you sleepy.

Is there a cause and effect?

I decided to ask Dr. Roland Powell, my pet MD, who told me that he actually has had (female) patients complain to him about this.

Why doesn't he just tell them to "get a life"?

"Then they wouldn't pay me."

'Nuff said. So, what causes men's sleepiness after sex?

"Well, there's a lot in what you say about having sex just before you go to sleep and therefore feeling sleepy—especially if there's been some exertion to help to release natural endorphins that relax us.

"You'll also find that, just before and during orgasm, you experience a quicker heart rate and higher blood pressure. After orgasm, the heart rate slows and the blood pressure drops, and both of those are conducive to sleep.

"But that's not all. An orgasm—male or female—also releases oxytocin, which is the same hormone that's released when women are breastfeeding and it's designed to put babies to sleep. Obviously, it has the same effect on adults. Yes, and that means women too."

◎ Sex between snakes lasts between six and twelve hours.

Is It Possible to "Enlarge Your Penis"?

Consider the following joke:

A man is in his favorite (expensive) restaurant when he notices a beautiful woman sitting alone at a nearby table. He calls the waiter over and asks for their most expensive bottle of Champagne to be sent over to her. The waiter does this and the woman sends a note to the man, which reads: "For me to accept this bottle, you need to have a Mercedes in your garage, a million dollars in the bank, and seven inches in your pants." After reading her note, the man sends back a note of his own: "Just to let you know, I have a Rolls-Royce in my garage, twenty million dollars in the bank, but not even for a woman as beautiful as you would I cut off three inches. So, please send back the bottle."

OK, first off, here're the facts: according to Gebhard and Johnson (1979), the average erect penis of males in the U.S. is five to seven inches and the average circumference is four to six inches.

So, what can the man who is below average, or above average but desperately insecure, do? Alas, it turns out, not much at all.

If a chap's lost (almost all of) his penis in—say—a bacon-slicer moment, then there is the possibility of reconstructive surgery. However, this is an incredibly intricate and laborious medical procedure and is simply not appropriate for purely cosmetic purposes.

Then there are silicon rods, which are inserted to help men with erectile dysfunction, but a flaccid dick is not the same as a small one. There's really not a lot that can be done.

So, what is one to make of all the advice available on the Internet? I mean, not many days go by when I don't receive an invitation to enlarge my penis.

Patches, hypnosis, exercisers, pills, pumps, and something called an extender device (least said soonest groin-ache mended). Oh yes, and surgery.

But hang on a second, Mitch, didn't you just say that surgery wasn't—what was the word you used?—"appropriate"?

Yes, but that doesn't stop people from offering it and, presumably, people taking them up on it. Desperate people will do almost anything. And looking at some of the "before" pictures on such Web sites, I think we might have to redefine the meaning of the word "desperate." I have seen bigger penises on a toy soldier.

Mind you, even the "after" pictures—presumably touched up (if you get my drift)—aren't that impressive. Who could be bothered? Wouldn't it be cheaper—and less painful—to buy a fast car?

To be fair to the companies offering enlargement surgery—and there's no reason why one should, given that it could be argued that they're trading in (and taking advantage of) people's miserable insecurity—they can make a tiny difference by adding a little fat to increase the length or girth, but we are talking tiny (before and after).

Still, there's clearly a market out there. Put "increase the size of your penis" in inverted commas as a specific phrase into Google and just see how many hits you get. I got thirty-six thousand, and nearly all of those were ads.

Pills seem to be the preferred method—100% GUARANTEED!!!—and while they might have the teensiest effect on teensy todgers (surely not every claim is a lie?), their efficacy would seem to be limited—or else every under-endowed bloke would be taking them.

So, there you are, a man in possession of a miniscule cock, and you hand over your hard-earned to someone who will almost certainly give you hard evidence that there's no hope whatsoever for you. In other words, it's what you might call a zero-sum game.

Men Who Were Well-Endowed

Gary Cooper
Aristotle Onassis (*"the secret of my success"*)
Jack London
Frank Sinatra (*according to ex-wife Ava Gardner, "there's only ten pounds of Frank but there's one hundred and ten pounds of cock"*)
John Derek
Henri de Toulouse-Lautrec (*not just big in relation to his size but BIG anyway*)
Errol Flynn (*his party piece was to get it out and try to play "You Are My Sunshine" on the piano*)
Rasputin (*said to be thirteen inches long*)
Jimi Hendrix (*said to be as big as his guitar*)
Charlie Chaplin (*"the eighth wonder of the world"—at least, according to him*)
Humphrey Bogart
Marlon Brando

Babe Ruth

Milton Berle (*the comedian once won a bet when he was challenged by a well-hung man. Berle, however, only pulled out enough of his "manhood" to win the bet and no more)*

John Dillinger

Chico Marx

Joe DiMaggio

Men Who Were Not Well-Endowed

Cary Grant (*according to ex-lover Maureen Donaldson*)

King Edward VIII (*"the smallest pecker I have ever seen," according to a "friend"*)

Napoleon (*just one inch when he died*)

Ernest Hemingway (*"the size of a .33 shell"*)

F. Scott Fitzgerald (*he compared his to Hemingway's and decided that they were both somewhat lacking in the manhood stakes*)

Vaslav Nijinsky

> ◎ Vaginas are, on average, between three and four inches long and they expand during intercourse. Including her clitoris, vaginal lips, and internal spongy tissue, a woman has just about as much erectile tissue as a man, but most of it is inside her body.

What's the Best/Worst Pick-Up Line?

Regular readers of my books—and a big hello to you both—might remember that I ran a list of chat-up lines in *This Book*.

Here are some of them:

"You don't know me but I dreamt about you last night."

"Did it hurt when you fell from heaven?"

"If I could rearrange the alphabet, I'd put U and I together."

"Can I buy you a drink or do you just want the money?"

"Do you believe in love at first sight or shall I walk past you again?"

"How do you like your eggs? Fertilized?"

"Get your coat, love, you've pulled."

"Is that a ladder in your tights or the stairway to heaven?"

"This body leaves in five minutes, be on it."

"Your clothes would look great on my bedroom floor."

"Can I have your picture so I can show Santa what I want for Christmas?"

"I may not be the best-looking bloke in the room but I'm the only one talking to you."

"You're a thief: you've stolen my heart!"

"Can I borrow your mobile? I told my parents I'd phone them when I met the girl of my dreams."

"Is it me or do you always look this good?"

"The word 'beautiful' wouldn't be the same without U."

In 2004, a British brewery ran a competition, which was won by the line "Well here I am! What are your other two wishes?"

Meanwhile, in 2005, psychologists from Edinburgh and Lancashire universities addressed this vital subject.

They didn't particularly rate direct requests for sex but thought those were better than the desperately corny "Do you have space in your bag for my BMW keys?" and "You're the star that completes the constellation of my existence." Ouch!

According to Dr. Christopher Bale, who led the research, "The highest-rated lines were those reflecting the man's control of a situation, wealth, culture, and spontaneous wit."

The scientists reckoned that examples of good lines were to say "Ten-ton polar bear" and then, when you get a puzzled reply, retort, "Well, it breaks the ice." Another good idea is to steer the conversation toward music in order to say, "the 'Moonlight Sonata' or, to give it its true name, *'Sonata quasi una fantasia'*: a fittingly beautiful piece for a beautiful lady." Sounds like the sort of thing you'd expect to hear from an Italian head waiter.

A better line was "There's something in your eye. Oh, it's a sparkle."

The sad, prosaic truth is that if a bloke's good-looking enough, then (almost) anything he says will work and, conversely, that if he's an ugly bastard then (almost) anything he says won't work.

The same is true for women.

- In ancient Greece, tossing an apple to a girl was a proposal of marriage. Catching it meant she accepted.
- The liver, not the heart, is the sign of romance in northern Morocco. When a Moroccan girl falls in love, she says, "Darling, you have stolen my liver."
- 6 percent of American men propose marriage by phone.
- Male western fence lizards do push-ups on tree limbs as a courtship display for females.

Is It True That Royals Are Watched by Flunkies While They Have Sex?

I phoned Buckingham Palace. No, honestly, I did. The first thing you hear is a recorded voice saying, "If you've been told to phone this number as a matter of urgency, then please hang up, as it's a hoax" (or words to that effect). I got through to the press office and told an utterly charming woman that I was Mitchell Symons, the author of *Why Girls Can't Throw*—which prompted the usual five-minute conversation (useful as an ice-breaker). Then I asked my question—qualifying it as a real "when did you stop beating your wife" question.

Lots of giggles.

"Never, surely!"

Well, obviously I knew that it wouldn't still be the case, but did it ever happen?

"You'd have to ask an historian."

OK. But what about people having to witness royal births? Does that still happen?

She went away and came back a minute or so later. "Princess Alexandra's birth in 1936 was the last occasion when a home secretary was required to be present."

That's pretty recent.

She agreed and went on to explain that until the reign of Queen Victoria, all sorts of lords, ladies, and busybodies (my word, not hers) had to attend royal births—albeit in the room next door. Victoria cut it down to just one cabinet minister—the home secretary—and so it continued until Christmas Day 1936, when Princess Alexandra was born.

King George VI then said that no one not directly in line to the throne would have to have their birth witnessed, and when, in 1948, a direct heir (Prince Charles) was born, King George VI decided that this arcane practice was "neither a statutory requirement nor a constitutional necessity."

As for the flunkies watching royals actually having sex, according to historian Dr. Michael Morgan, "Not since medieval times and not always then. What tended to happen was that the newly married couple would be ceremoniously put to bed by their guests while the marriage bed would be blessed by a priest. Then the couple would consummate their marriage—usually unattended. They were expected to employ the missionary position—as it subsequently came to be known—because that was thought to be the only position to use to create boys. And boys were, of course, important for dynastic reasons."

No One Wears Chastity Belts Anymore . . . *Do They*?

I take it that you know what a chastity belt is? If not, I should explain that it's a locking device that prevents the wearer from engaging in sexual intercourse. In order to qualify as a chastity belt, it must be lockable.

These belts were first used in the Middle Ages, to prevent women from being unfaithful and/or raped while their husbands were away at the Crusades.

So, do they still exist? You bet!

There's a Sheffield company named Tollyboy ("Protecting your assets since 1956"), which runs a thriving business on the

> ◎ The first couple to be shown in bed together on U.S. prime-time TV were Fred and Wilma Flintstone.

Internet, selling female and male chastity belts/anti-rape devices and ankle, wrist, and biceps bracelets and neck collars, as well as other products, like plugs and restraints.

The company was founded by Hal Higginbottom, an engineer at Sheffield University, who was challenged by a lecturer to make a modern-day chastity belt. When Higginbottom died in 1998, his apprentice Richard Davies took over the business, and it was he who told me that "70 percent of orders are from men."

To protect women or to stop them having fun?

"No, they are for men. Orders would run 70–75 percent male belts, the female belt is a rarity. Belts are used as an aphrodisiac, human nature being as it is, people want what they can't have."

Was he satisfied that no one's actually being imprisoned in them?

"I know of no one who is imprisoned in one, but that would be some people's fantasy. In any case, the police, fire brigade, or any decent engineering firm could cut one off, so to keep someone imprisoned in one would require keeping them away from the outside world, phone or e-mail, etc."

How much does a typical one cost and how many does he sell in a year?

"Three hundred pounds ($590.79) and wildly variable, but not enough to make much money."

What Do They Call a Brazilian Bikini Wax in Brazil?

First of all, I should explain precisely what a "Brazilian" is. It's a method of hair removal which leaves a woman with only the tiniest amount of pubic hair (otherwise known as a "landing strip"). The reason it's called "Brazilian" is because the practice started in Brazil. During the 1990s, it was popularized in New York by the Brazilian beauticians, the J Sisters, whose salon boasts an autographed photograph of Gwyneth Paltrow, with the inscription: "Thanks, you changed my life."

It's lucky for me that I lack the embarrassment gene, because I have to tell you that even I found myself turning red as I phoned up all sorts of Brazilian organizations to ask them this incredibly daft question. True, most of them failed to understand what the hell I was talking about, but I think I managed to jeopardize Anglo-Brazilian relations for at least a generation.

Eventually, a very nice lady at somewhere called Brazilian Contemporary Arts agreed to help me. "We just call it 'depilar,'" she said, obviously marking me down as a true British eccentric.

Not a Colombian or a Peruvian or, indeed, a German?

"No, just 'depilar.'"

But you're aware that the rest of the world calls it a Brazilian?

"Yes."

And what do you think of that?

Silence.

Hello.

"Is that all you want?"

I suppose so. Thank you.

I guess it was a dumb question, but underlying it is something really rather important. What kind of society do we live in, where women are told that shaving their pudenda to a little "landing strip"—or, indeed, rendering them totally bald—is the fashionable thing to do? Isn't it a teensy bit perverted for grown-up women to attempt to turn themselves into prepubescent girls?

And what kind of man finds a bald pussy sexy? And why would any woman want to hook up with the sort of man who finds a bald pussy sexy?

You get my point.

At this juncture, I should say that, when it comes to pubic hair, I'm very much a meat-and-two-veg man. Although I think that the completely shaven look is (unacceptably) pervy, I find the opposite extreme just as unpalatable. Put it this way, I'm a suburban man and I like pubic hair to be like a suburban garden—neither wild jungle nor bald paving stones: merely properly trimmed and well-maintained. A neat verge—is that too much to ask?

Indeed, so conservative am I in this area, that I really think a short back and sides (and shampoo) is all that any self-respecting woman should go in for.

Women Who Dyed Their Pubic Hair

Jayne Mansfield (*blond*)
Carole Lombard (*blond*)
Jean Harlow (*platinum blond*)

Are the Letters Sent to "Men's" Magazines Genuine?

You know the sort of thing: "Dear Sir, I always thought that the letters in your magazine were made up until I had an experience which I would like to share with your readers. I was traveling on this train when this fantastic-looking girl got on. To my delight, she sat directly opposite me. To my amazement, all the other passengers got off at the next stop. To my . . ." Etc., etc.

As I've already mentioned, when it comes to men's magazines, I've got some experience. Some twenty years ago, I used to contribute—no, not letters—articles to *Penthouse*. I—or rather my alter ego, Mike Hunter (try saying it a few times and you'll understand the choice of pseudonym)—wrote things like "The *Penthouse* Guide to Hen Nights," "The *Penthouse* Guide to Stag Nights," "Love on the Dole," and "Convent Girls." The aim of the articles was to inform and entertain, but never—as far as I was concerned—to arouse. I loved writing that sort of stuff and, indeed, I really do think that I did some of my best work for *Penthouse*. I never did find out whether the letters in that magazine were true or false. My guess is that they were based on readers' letters but were then embellished/cleaned up/tidied up as necessary.

So, I phoned around some of the other guys I know who've worked for "men's magazines," and asked them for their opinion. The consensus was that almost all of the letters were indeed made up but that some were at least based on readers' true experiences. The chances are, though, that the readers' letters were likely to be the least interesting, or, as one fellow—a veteran of jazz mags—put it, "If they give you a boner, they're made up."

Apart from anything, it's safer for the staff of gentlemen's magazines to write their own letters' pages, as it means that they won't be the victims of spoofs. One such magazine once found itself publishing a series of letters on the joys of sex with amputees. After a few months, it became obvious that the whole thing was a practical joke, and, you might think fittingly, the magazine didn't have a leg to stand on.

One in Ten Men? Surely Not?

Does it even matter? As one gay friend put it when I asked him, "I don't care what percentage of the male population is gay so long as the gay in my bed is 100 percent!"

I think that it does matter, because it's become such an emotive, totemic figure (and, no, I'm not talking about the one in ten that UB40 were singing about).

Bearing in mind the usual health warnings about statistics ("There are three types of lies: lies, damn lies, and statistics"), this one would seem to have some kind of accuracy.

The one-in-ten goes back to Kinsey's original 1948 report. Since then, there have been many more investigations into the prevalence of homosexuality, and the findings have been fairly consistent—if slightly lower.

A 2006 survey in Britain produced the figure of one in sixteen.

In 1993, Janus and Janus (great names!) found that 4 percent of men considered themselves homosexual and 5 percent bisexual—which adds up to 9 percent. The equivalent figures for women were 2 and 3 percent.

Much more recently, in 2005, Mosher, Chandra and Jones found that 90 percent of men considered themselves to be heterosexual. Now, that obviously leaves 10 percent, or "one in ten."

Unfortunately, it's a little more complicated than that, because the other 10 percent variously described themselves as homosexual, bisexual, and "something else" (the mind boggles).

Interestingly, the figures were almost identical for women.

So, what can we conclude?

I think we can safely say that "one in ten" covers men who are not heterosexual. Beyond that, who can tell what people get up to in the privacy of their homes? Who can vouch for the accuracy and honesty of what respondents tell scientists and pollsters? Some men who are gay refuse to acknowledge it even to themselves.

There's an extra dimension, too: during the writing of this book, I've learned just how fluid categories can be—and, therefore, how worthless categorization can be. What about the otherwise heterosexual man (OHM) who has sexual thoughts about men? Or the OHM who has anal sex with a postoperative transsexual woman? Or my friend (whose name I will not divulge—or at least not while he continues to pay me) who, although an OHM, likes nothing better than for his girlfriend to strap on a dildo and give him one up the bum?

And who knows where people go to when they're alone in their beds?

> ◎ Homosexuality remained on the American Psychiatric Association's list of mental illnesses until 1973.

Men You Might Not Have Known Were Gay

Vaslav Nijinsky
Molière
Tennessee Williams

Brian Epstein
Dag Hammarskjöld
John Schlesinger
King James I
Sandro Botticelli
Socrates
John Maynard Keynes
Charles Laughton
Edward Albee
James Baldwin
Henry James
Ralph Waldo Emerson
Frederick the Great
Sophocles
Franz Schubert
Thornton Wilder
Maurice Ravel
Ludwig Wittgenstein
James Coco
President James Buchanan
Lorenz Hart
Raymond Burr
Leonardo da Vinci
Clifton Webb
Walt Whitman
George Cukor
Monty Woolley
Cole Porter
Peter Tchaikovsky
Plato
Vincente Minnelli
William S. Burroughs

Do Nuns and Priests Have to Be Virgins?

No, I'm not stupid: I know that nuns and priests are tied to their vows of chastity. What I wasn't so sure about, however, was whether a nonvirgin was barred from the priesthood or from entering a convent.

So, I phoned Father Paul Embury, National Officer Vocation (responsible for the formation of priests in Britain).

"No," said Father Paul, "priests don't have to be virgins. When any candidate presents themselves, we look at every facet of their lives. We want to know their sexual history—not just physical sex. If somebody has been sexually active in the past, we'd want to see there'd been a period of abstinence and we'd want to know that they were orienting themselves toward a life of celibacy. Even someone who's in a relationship that's nonsexual, we'd want to know that they've put some distance between themselves and their girlfriend before they can commence with their formation and training.

"There's also a psychological assessment which would take account of such issues."

We've been talking about priests—do you happen to know if the same is true for nuns?

"Yes, it's the same."

But then afterwards, of course, they have to be totally celibate.

"Yes, though interestingly, if a priest falls from grace, he's not automatically dismissed—although he would be expected to take responsibility for any child he fathered.*

"There is also a slight difference between, on the one hand, monks and nuns and, on the other, priests. Monks and nuns take solemn vows of chastity, poverty, and obedience. Whereas diocesan priests take a promise of obedience and a promise of celibacy: these are promises and not solemn vows. This is a nuance that not everyone would get. The difference being that a vow is binding whereas a promise isn't."

Why wouldn't a priest make such a vow?

"Monks and nuns vow to remain celibate for the rest of their lives whereas, with priests, the pope might change the stance on married priests, and then they wouldn't have to be celibate. And then there are the Anglican ministers, who became priests after leaving their church over the ordination of women: they're not expected to be celibate."

Men Who Considered Becoming Priests

Tom Cruise (*at age fourteen, he enrolled in a seminary but dropped out after a year*)
John Woo (*the film director trained as a priest*)
Pete Postlethwaite
Joseph Stalin

* After our conversation, I chanced upon a newspaper article about a seventy-three-year-old Irish priest, who had fathered a child by a thirty-one-year-old woman and not only had he not been chucked out of his job but he'd been treated with sympathy by all around him. This amply echoes Father Paul's point..

Christopher Marlowe
Ben Vereen
Alan Bennett
Charles Darwin
Bob Guccione
Bernhard Langer
Gabriel Byrne
Roberto Benigni

Women Who Might Have Become Nuns

Heather Graham (*her parents had ambitions for her to become one*)
Kristin Scott Thomas (*at sixteen, she enrolled in a convent school with a view to becoming a nun*)
Cher (*"There was a time when I nearly became a nun myself, but it didn't last long"*)

People Who Died Virgins

Immanuel Kant
Sir J. M. Barrie
Nikolai Gogol
Queen Elizabeth I
Sir Isaac Newton
Anton Bruckner (*despite his fixations with teenage girls*)

What Is the Greatest Male Sexual Lie?

Well, for a start, it might just be that you can't get pregnant if you have sex standing up (see chapter "Can You Get Pregnant If You Have Sex Standing Up?")—especially if uttered in order to persuade a girl to have unprotected sex.

Other than that, it's hard to know where to start—given that a man in possession of a stiffie is prepared to say pretty well almost anything for an invitation (note the juxtaposition of "stiffie" and "invitation"—rather clever, don't you think?).

Male sexual lies can be neatly divided into two distinct categories. Those told *before* sex (i.e., in pursuit of it) and those told *after* sex (i.e., to get away—bearing in mind the different male/female sexual imperatives discussed previously).

Before

I'll still respect you in the morning
This is *it*
My partner and I have an open relationship
God, you're so beautiful

I won't tell a *soul* about this (*used before with the stress on the fifth word*)
There's no one else but you
I'll only put it in a little way
I never knew it could be like this
I want to wake up next to you in the morning
I promise I won't come in your mouth

After

Of course I still respect you
I won't tell a soul about this (*used after with the stress on the first word*)
There's no one else but you
You're really very attractive
I tried phoning you but your number was constantly busy
Thank God you phoned—I lost your telephone number

What Is the Greatest Female Sexual Lie?

When it comes to telling the truth before, during, and after sex, women are, by far, the more honest gender. Or at least so they tell us, and so we chaps like to think. But I suspect it's true, because women are not only encouraged to be more in touch with their feelings but also to share them.

Having said that, they too are capable of mendacity. Here are some examples:

This is my very first one-night stand
My boyfriend and I are in the process of breaking up
You're the *best*
I've never done that with a man before

And which is the greatest female lie?

I think we all know that, fellows, don't we? Yup, that's right, it's "**Of course I'll still do that when we're married.**"

Why Is the Most Common Position for Sexual Intercourse Called the Missionary Position?

On all things linguistic, I am obliged (through sheer indolence more than anything else) to consult Dr. Caron Landy.

"Ah, you mean *venus observa*."

Do I?

"That is its technical name."

So, why is it called the missionary position? Is it because this was the way in which missionaries showed the natives the "proper" way to have sex?

"A common misconception, I'm afraid."

Tee hee!

"What?"

Misconception. Good pun.

"If you say so," she said glacially. "In any event, the missionary position, in which the man lies on top of the woman, is a relatively recent expression, dating from after World War II."

I was surprised. As recently as that?

"The expression, that is, not the position. Yes. It seems that missionaries didn't, in fact, express any opinion on how sex should be performed—why on earth would they? However, that didn't stop mischievous Pacific Islanders from including them in their range of impersonations of how different people had sex."

That's extraordinary!

"It is, isn't it? Of course the islanders themselves were more inclined to have sex with the woman on her hands and knees . . ."

Doggie-style . . .

". . . quite, and they decided to characterize the missionaries as being so straitlaced that they would only do it in—what the islanders considered to be—the least embarrassing fashion possible. From there, in what might be termed an intriguing self-fulfilling prophecy, the expression "missionary position" came to be the default description of the sexual position of a man lying on top of a woman."

Things People Do after Sex

Cuddle
Go to the bathroom
Eat
Drink
Go to sleep
Watch TV
Read
Talk
Argue
. . . it again

◎ More Americans lose their virginity in June than any other month.

Are There Any Countries in Which Homosexuality Is Still Illegal?

Yes.

Perhaps surprisingly, it's only been completely legal in the so-called enlightened West in the past few years—it was still illegal in Ireland until 1993.

The countries where it's still illegal are, alas, the usual suspects: Muslim countries and African countries (though not South Africa).

As most people know, Sharia law forbids homosexuality. In Africa, at the last count, male homosexuality was illegal in twenty-nine countries and female homosexuality was illegal in twenty.

It is also still illegal in India.

Premature Ejaculation—Doesn't It Just Save Time?

Premature ejaculation is not uncommon. Google the two words and you get more than two million hits—many of them offering advice to sufferers.

So, why is it shrouded in so much furtiveness? To give you an idea, these were the only famous PEs I could find: King Edward VIII, John Maynard Keynes, Tony Curtis (in his youth), Havelock Ellis (until the age of sixty and without ever achieving vaginal penetration), Johann von Goethe (until his late thirties), Groucho Marx (all his life), Henry Fonda (according to ex-wife Margaret Sullavan), Rock Hudson (according to a woman he had sex with—now he probably did just want to save time) and John F. Kennedy (another man who would have wanted to save time—but only because he had so many more women to attend to).

Not a huge haul, is it? And that was after a truly epic trawl through all my extensive library.

As you'll see later, in the answer to the question "Is It Possible for Men to Fake Orgasm?", I used to suffer from the opposite problem (no, not immature ejaculation), so I used to find it hard to sympathize or empathize much with chaps who were able to do it quicker than I could.

But after one member of my teenage crowd was "outed" by a particularly cruel girl as a premature ejaculator, I realized that the inability to consummate a relationship for more than a few seconds could indeed be seen as an affliction.

Generations of Zen Buddhists have spent their lives contemplating the question "Does the tree that falls in the forest make a sound if no one is there to hear it?" Me and my mates were a little more prosaic: can a man be said to have had sex if the girl hasn't even noticed it?

Cruel, I know; but this was the 1970s and we didn't know what consciousness was, let alone thought about raising it.

I never spoke to the poor lad about his problem. Indeed, as I recall, he sort of disappeared from the scene pretty soon afterwards.

The cruel girl married a real estate agent. She deserved nothing better.

As I canvassed my (current) friends, only one of them admitted to being a premature ejaculator—and, even then, only in his youth. Roger (for want of a better pseudonym) told me that, in his case, it was a mixture of nerves and lust. "Just the thought that I was going to, you know, get it away was enough to set me off. Jeez, you spend all that time tugging away at yourself just at the thought of having sex with a 'real woman.' Then, when the time comes, you just explode as soon as you're about to start."

I told him that I could understand that happening the first time—big build-up, etc.—but surely not on subsequent occasions?

"Are you kidding? Those were even worse! Now the nerves started kicking in. You know, 'Oh God, I hope I don't come too soon!' And of course you do: it's a self-fulfilling prophecy."

How did you cure it?

He shrugged his middle-aged shoulders. "Time heals most things. With enough repetition, you get less lustful and less nervous. Also Tina (his wife) helped enormously by not making a fuss of it when we were first together."

Bet she wishes you were a little more premature these days.

There was a flash of youthful anger in his eyes. "No, Mitch: to answer your original question: when it comes to sex, not all people want to 'save time.' Don't judge everyone's sex life by your own."

Ouch.

- Many species of bird copulate in the air. In general, a couple will fly to a very high altitude, and then drop. During their descent, the birds mate.
- Minks have sex sessions that last, on average, eight hours.

Sexual Offers That Were Turned Down

Marianne Faithfull turned down Bob Dylan and Jimi Hendrix
Marlon Brando turned down Tallulah Bankhead
Olivia de Havilland turned down Leslie Howard
Marlene Dietrich turned down Ernest Hemingway and Adolf Hitler (*after approaches to her were made by Goebbels and Ribbentrop on his behalf*)

Ava Gardner turned down Howard Hughes and George C. Scott
Mary Pickford turned down Clark Gable
Greta Garbo turned down Aristotle Onassis
Katharine Hepburn turned down Douglas Fairbanks Jr. and John Barrymore
Clara Bow turned down Al Jolson
Zizi Jeanmaire turned down Howard Hughes (*this was after he'd bought the entire ballet company for a film—purely to seduce her*)
Gertrude Stein turned down Ernest Hemingway
Jaclyn Smith turned down Warren Beatty
Joan Collins turned down Richard Burton
Bette Davis turned down Barbara Stanwyck
Bette Davis turned down Joan Crawford (*according to Hollywood legend—although their lifelong enmity is equally likely to be due to the fact that Davis had an affair with Franchot Tone while he was married to Crawford*)
Marilyn Monroe turned down Joan Crawford
Anthony Perkins turned down Ingrid Bergman
Jacqueline Susann turned down Coco Chanel
Grace Kelly turned down Bing Crosby
John Wayne turned down Marlene Dietrich

What's So Big (and Clever) About Bigamy?

What indeed?

Bigamy is one of those crimes that carries its own punishment—two wives and two mothers-in-law. Mind you, Oscar Wilde—not the most uxorious of men—reckoned, "Bigamy is having one wife too many. Monogamy is the same." Though to be fair to Oscar, I've also seen the same quote attributed to Erica Jong and Gloria Swanson—both of whom I met, by the way (just thought I'd mention it).

Interestingly, there are very few instances of women committing bigamy. This isn't (just) because women are far too sensible to sign up for more than one husband—at least not concurrently—but because, generally speaking, women aren't as proficient as men at compartmentalization.

Bigamy has always been uncommon but it's particularly rare nowadays—partly because of better communications (especially mobile phones) and also because men don't have to marry women just to get a leg over, so there's less incentive.

Indeed, in my researches, I found very few "famous" bigamists. Rudolph Valentino, George Gissing, Jerry Lee Lewis (at age sixteen!), Anaïs Nin, Judy Garland, and Steven Seagal were all wittingly or unwittingly married to two people simultaneously. It is also said that John F. Kennedy married a socialite named Durie Malcolm in 1939 and was still married to her in 1953 when he married Jacqueline Bouvier. Meanwhile, Sir Michael Redgrave and Lord Jeffrey Archer both had bigamous fathers.

- In Hungary, bigamous men were once punished by being compelled to "live with both wives simultaneously in the same house."
- Married men change their underwear twice as often as single men.
- Percentage of American men who say they would marry the same woman if they had to do it all over again: 80. Percentage of American women who say they would marry the same man: 50. Percentage of American men who say they are happier after their divorce or separation: 58. Percentage of American women who say they are happier: 85.

Why Do Men—Especially Men in Bars—Assume That the More Desirable a Woman Is, the More She'll Desire Them?

You know what I mean. A group of guys are gathered together in a bar. A beautiful woman walks in and the conversation goes something like this:

"Hey! Did you see that?"

"Wow!"

"She's gorgeous!"

"Bet she loves it!"

"She's hot for it!"

"Dirty slut!"

All right, you get the drift. The woman has done nothing more provocative than walk into a bar in possession of a beautiful face.

So, what's it all about?

I decided to consult the eminent psychiatrist Anne Kreeger.

"What do *you* think it's all about?"

I'm sorry, I came here for answers, not questions.

"Think about it, Mitch," she said, crossing and uncrossing her (really rather nice) legs in a way that reminded me of Dr. Jennifer Melfi in *The Sopranos*.

I don't know, maybe she gives off some sort of signal; maybe her beauty is predicated on that.

"No!" She made me feel like a particularly stupid student. She saw my face fall. "Well, perhaps, but no that's not it. They're projecting their desires onto her. This is something that men, particularly, are wont to do. Look at Hitler . . ."

Do I have to?

"No, seriously. That which he hated in himself—particularly the sexually aberrant behavior—he projected onto the Jews, whom he accused of performing the most appalling acts of sexual depravity. So it is here. The men find this woman beautiful and sexy and so they choose to believe that she's 'gagging for it.'"

Psychological term, is that, Doc?

"What?"

Gagging for it.

"You'd be surprised the language we use when we're not working. But it's interesting that you contextualize this encounter with a group in a pub—thereby adding to the situation the two key factors of peer group pressure and inebriation."

Do Celibates Get More Work Done Than the Rest of Us?

I see no more action than any other happily married middle-aged man, but by no stretch of the imagination could I be described as a celibate.

And it's not only the (limited) time I spend in blissful sexual congress that I'm obliged to consider but also the (not so limited) time I spend thinking about sex.

All right, so I also get to work, eat, watch sports on TV, and play Internet poker—none of which involve sex (except, at the moment, writing this book), but it's rare that my libido is disengaged for long.

That's how it goes.

I have an acquaintance who once mentioned to me that he and his partner didn't go in for sex anymore.

So, I phoned him and told him I was writing a book and could he help me with this question. And guess what? He demanded anonymity! I can't think why . . .

Anyway, Vivien, as I shall call him, confirmed that he is, in fact, (effectively) celibate. "I have such a low sex drive—always have had—that it seemed easier to desist altogether rather than leaving it up in the air."

What about your partner?

"She's fine with it. Truth be told, she was never much up for it in the first place. Sex was never important to our relationship—except in the breach: that's to say, at different times, we each felt guilty for not doing more about it. The day we decided simply to stop having sex was the best day in our relationship."

So, do you get more work done than noncelibates?

He chuckled. "I do, but it's not quite for the reason that you think. Look, let's say that the average guy of our age has sex once a week and that that lasts for, say, twenty minutes."

OK, but that would include the foreplay and the obligatory cuddle afterwards.

"Of course! But seriously, Mitch, twenty minutes is nothing out of a week containing several thousand minutes. How much more work would you be able to do at that time—always assuming that you would do any work at eleven o'clock on a Saturday night? No, the big difference between me and you isn't that you have sex and I don't; it's that you think about it and I don't. *That's* where I save time. I don't fuck secretaries . . ."

Nor do I!

"I don't have affairs . . ."

Nor do I!

"I'm not led around by my dick . . ."

Ah, you've got me there.

"So, all that time you spend thinking about it, talking about it, wondering about it, worrying that you're not getting as much as you think you ought to be, looking it up on the Internet—all that time you could be working. So yes, on that basis, I'd say that celibates do get more work done than sad sacks like you."

I am NOT a sad sack. Am I?

Or is to ask the question to answer it?

Oh dear.

"Sex is one of the nine reasons for reincarnation—the other eight are unimportant."

(HENRY MILLER)

"Sex has never interested me much. I don't understand how people can waste so much time over sex. Sex is for kids, for movies—it's a great bore."

(SIR ALFRED HITCHCOCK)

People Who Never Married

Sir Edward Heath
Greta Garbo
Sir Isaac Newton
Florence Nightingale
Ludwig van Beethoven
Frederic Chopin
Queen Elizabeth I
Henri de Toulouse-Lautrec
Jane Austen
Louisa May Alcott
Giacomo Casanova
Lillian Gish
George Gershwin
Jean-Paul Sartre
René Descartes
Immanuel Kant
Friedrich Nietzsche
Patricia Highsmith
Benny Hill

How Devoted to Watersports Do You Have to Be to Do It After Eating Asparagus?

In *Why Girls Can't Throw*, I tried to answer the question of why it was that asparagus makes our wee-wee smell (turns out that it doesn't affect everyone, but for those of us who have this gene, there's a sulfurous compound found in foods like asparagus and garlic, which, when broken down by our digestive systems, comes out in our urine half an hour later. And it smells rank).

Anyway, the point I'm making is that although I'm prepared to accept that there are people who enjoy peeing on each other—it's not called "water sports" for nothing—might there not be a limit to aficionados' enthusiasm?

It was time to call my mate Rick.

"Oh yes," he said cheerily enough, "I'm definitely on for water sports. Though not for hard sports . . ."

Hard sports? That's a new one on me.

"Poo-poos."

Oh no, that's disgusting!

"Why? If that's what consenting adults want to do?"

Because . . . because . . .

"You are so judgmental! I don't do hard sports because it doesn't appeal to me, but if it did or if I were with a woman who really really wanted me to . . ."

Enough! Go back to water sports!

"All right. Having a really sexy girl pissing all over you in a golden shower can be just fabulous . . ."

. . . I'll take your word for it.

"Yes, you will, won't you? And if she wants me to do it to her, why not? It's only water, after all."

Aha, but what if one of you has been eating asparagus?

"I hear what you're saying but in the moment—I mean, at the precise moment in time that you're doing it, you just don't care: it simply doesn't matter. That's the whole point of great sex: you suspend your normal fastidiousness and just go for it.

"Except, of course, *you* don't, do you?"

I thanked him for his contribution and reflected on what he'd said. And you know what? I hear what *he's* saying, and I know I'm old-fashioned, but there are some things that really are a step too far, and urine smelling of asparagus is one of 'em.

◎ The average person produces about twelve thousand gallons of urine over the course of their lifetime.

Why Do Women on HRT Always Look As If They've Just Emerged from a Tanning Parlor?

I am aware that many of the questions in this book are not necessarily on the tip of every tongue. This, for example, is obviously quite a recherché question (if recherché means what I think it means). Chances are you've never even stopped to think about such a question—let alone lain awake at night pondering it. But grant me this at least: whenever a female celebrity of a certain age appears in a magazine or goes on TV to extol the virtues of hormone replacement therapy (HRT), they do seem to have the most extraordinary suntans. In fact, I'd go so far as to say that they're almost orange.

Why?

Is there a link or is it just coincidence? Or am I plain wrong?

I wouldn't even begin to know how to answer such a question, but Mrs. Symons, a fine woman, as I think I've already told you, is approaching the sort of age when she might just benefit from a little knowledge on the subject, so I asked her to report from the front line.

She accepted the assignment and, more to the point, accepted the premise behind the question.

After an exhaustive study—which, for some reason, seemed to necessitate an awful lot of expensive girlie lunches—she told me that there is a sort of cause and effect but that it's more cosmetic than medical.

"Before a woman goes on HRT, she's usually feeling pretty rundown. That's why she takes it."

Sounds fair enough.

"Once she takes HRT, she will often feel an enormous improvement in her general sense of well-being and in her sex life."

Why would that make her skin change color?

"Wait for it, I'm getting there. Feeling good about herself makes her want to feel even better about herself. Suddenly, she's taking additional pride in her appearance. Hence the tan—real or fake. She wants to look her best."

Even at the risk of turning orange?

"Even at the risk of turning orange.

"By the way, you should know that this is a huge industry. Statistics show that between 20 and 50 percent of women in the Western world who are between the ages of forty-five and seventy have taken or are now taking HRT."

How does it work?

"HRT controls hot flushes and night sweats and relieves vaginal soreness due to dryness. It also reduces hip fractures caused by osteoporosis or 'crumbling bones.'"

OK. And it doesn't make you turn orange?

"Not necessarily."

Suddenly, I needed reassurance. "*You're* not going to turn orange, are you?"

But answer came there none, and I now live in almost perpetual dread that one day I will end up with a wife who's half-woman/half-orange.

Is It Possible to Die of a Broken Heart?

The answer to this question illustrates a wider point about mankind: our eagerness to distance ourselves from our emotions.

Let me explain. Hundreds of years ago, grief was accepted as a perfectly valid cause of death. Then, "reason" and "enlightenment" taught us to treat such a proposition with scorn, and it is only recently that the medical establishment have looked at cases of sudden cardiac arrests precipitated by strong emotional distress and concluded that it is indeed possible to die of a broken heart.

Tell that to Johnny Cash—except you can't, because he died of a broken heart just four months after the death of his wife, June.

However, notwithstanding J.C., the type of heart attack we're talking about—when an otherwise healthy person suffers from such intense stress that they die (the technical name for it is "stress cardiomyopathy")—is much more common for women than it is for men.

The reason for this could be prosaic: namely, that men are much more likely to die before their wives and therefore there are many more brokenhearted women than there are men. Though having said that, it should be pointed out that it is not only death that can precipitate the deadly release of the stress hormones that can stun the heart, but also traumatic breakups and shocks. So, perhaps we have to think again why women are so much more likely to be affected. Scientists have been reduced to surmising that it might have "something to do with their hormones" or the way in which "their brains are wired to their hearts." In other words, who can say?

A study was conducted by the Johns Hopkins School of Medicine, which looked at nineteen patients (all but one women) who had what, at first, seemed to be traditional heart attacks between 1999 and 2003. All the patients had experienced sudden emotional stress (news of a death, being present during an armed robbery, involvement in a car accident, even finding themselves at their own surprise party). The important thing was that although most were quite elderly, none had a history of heart problems.

When these people were compared with people who had had more "conventional" heart attacks, they found that the nineteen patients' arteries were healthy and unclogged but that the levels of stress hormones (such as adrenaline) in their blood were two to three times as high as in the "conventional" heart attack victims, and seven or more times higher than normal.

According to Dr. Ilan Wittstein of Johns Hopkins, "Our hypothesis is that massive amounts of these stress hormones can go right to the heart and produce a stunning of the heart muscle that causes this temporary dysfunction resembling a heart attack. It doesn't kill the heart muscle like a typical heart attack, but it renders it helpless."

Although people can—and do—make a full recovery from stress cardiomyopathy, there is obviously a serious problem with those victims whose original attack was precipitated by the loss of a loved one (as opposed to a shock), because although they might recover physically, there is less incentive for them to do so.

Geese often mate for life, and can pine to death at the loss of their mate.

Do Bisexuals—Page Woody Allen—Have Double the Chance of a Date on a Saturday Night?

"**Bloody greedy** bastards," said my buddy Pete, somewhat surprisingly. "They don't want to make up their minds which side they're on."

Are you kidding?

"All right, you got me—but I do feel a teensy bit pissed off with guys who are switch-hitters."

Why?

"Look, if they go with other men, they're gay, right? Nothing wrong with that: leaves more for the rest of us. But if they also play for our side, then they're using up valuable resources."

I pointed out that he seemed to have no problem with female bisexuals.

"On the contrary, the more the merrier: there's nothing I like better than a pair of carpet-munchers getting it on."

It's only when I'm with Neanderthals like Pete that I realize just how high my consciousness has been raised over the years.

Fortunately, I also have access to intelligent people—like the social anthropologist Dr. Lorraine Mackintosh.

"You must remember," she said, "that it's only in (relatively) recent times that we've categorized people at all. People just used to have sex with whoever—or, indeed, *whatever*—was at hand. Homosexuality, per se, is very much a nineteenth-century construct.

"Obviously, heterosexual sex has always been necessary for the propagation of the species, but when it comes to recreational sex, there have been ages that are less straitlaced than ours."

I'm surprised by that: I've always thought that mankind has become more permissive as it's evolved.

"Some and some. Social mores never follow a simple straight trajectory. And behavior can be extremely widespread and yet not be officially sanctioned by the society that, effectively, nurtures it."

All very helpful but it was time for me to find myself a living, breathing bisexual.

This wasn't as easy as it sounds. I know plenty of gays and lesbians but not—or so I assumed—any bisexuals.

In fact, it was only when I was asking my friend Hugh—whom I'd always taken to be exclusively gay—for help with finding someone who swings both ways that he revealed that he himself fits that bill.

"I didn't know that," I said, almost indignantly.

"There are lots of things you don't know about me. You know, Mitch, if you have a defect, it's that you tend to put people into boxes. Gay, straight, married, single. Life's more fluid than that. Sometimes it's more complicated; sometimes it's simpler. To quote the late James Dean, 'Why should I go through life with one hand tied behind my back?'"

Hang on, I spluttered, I've never seen you with a woman.

"Oh, I see, just because you don't know something, it doesn't happen. Does the tree that falls in the forest make a sound if . . ."

Don't get me started . . .

"No, the point I'm making is that it *does* make a sound—yeah, even if Mitchell Symons doesn't happen to witness it.

"So, yes, I'm a gay man but if I fancy a woman—and she fancies me—then I've no problem obliging myself. And her."

All right then, does being a bisexual double your chances . . . ?

"I'm not a 'bisexual.' Stop putting me in categories! I'm me—Hugh. Sometimes—usually—I have sex with men and sometimes, I have sex with women. And if I want to have sex with an animal, I would do that too. But that wouldn't make me a 'bestialist.' Any more than playing tennis once a month makes me a tennis player."

I didn't know you played tennis.

"The list of things you don't know is . . ."

I get it. Well, does being a bisexual—I mean, does being prepared to sleep with anyone . . . I mean, you know what I mean . . .

He put me out of my misery. "I very rarely spend an evening alone—except out of choice. But that's not because I'm gay, straight, or bisexual, but because I am totally gorgeous."

If you say so, Hugh.

People Who Were Bisexual

Nero
Michelangelo
Plato
William Shakespeare
Socrates

Queen Anne
Catherine the Great
Caligula
Pompey
Alexander the Great
Aristotle
Lord Byron
Peter the Great
Marie-Antoinette
Greta Garbo
Marlene Dietrich
Tyrone Power
Errol Flynn
Maurice Chevalier
Tallulah Bankhead
Janis Joplin
Judy Holliday
Salvador Dali
Siegfried Sassoon
Randolph Scott
Howard Hughes
Cary Grant

Lesbians Who Married

Billie Jean King
Virginia Woolf
Bessie Smith

Gay Men Who Married

Oscar Wilde
Peter Tchaikovsky

Sir Elton John
Raymond Burr
Vaslav Nijinsky
Cole Porter
Yukio Mishima
Charles Laughton
Malcolm Forbes
Vincente Minnelli
Sir Arthur C. Clarke

◎ Most giraffes are bisexual; so are most orchids.

Do Men *Really* Think About Sex Every Six Seconds?

As seldom as that, eh?

No, I'm only kidding—though when it comes to matters of sex, I do have a tendency to revert to my teenage self. But even then, I doubt if I thought about "it" more often than, ooh, twice an hour—tops. There are other things in life, you know. Why else did God invent food, sports, and television?

Precisely.

So, where does this "six seconds" idea come from?

Incredibly, nobody knows.

So, almost by default, we can conclude that it's a generic putdown by women—a line like "men are so pathetic, they think about sex all the time" evolves into "men are so pathetic, they think about sex all the time—every ten seconds, I've heard." Being more specific is a classic way of spreading a mistruth (see Marianne Faithfull and the Mars Bar—there was, as I explained in *Why Girls Can't Throw,* no confectionery in the Mick and Keef drugs bust but that didn't stop people saying that there was and then "proving" it by specifying the product). From ten seconds to six seconds is then simply a matter of embellishment.

For what it's worth, according to the Kinsey Institute, 54 percent of men think about sex every day or several times a day, 43 percent a few times per month or a few times per week, and 4 percent less than once a month. By comparison, 19 percent of women think about sex every day or several times a day, 67 percent a few times per month or a few times per week, and 14 percent less than once a month.

Six seconds indeed!

And how much time would that leave for Internet poker?

◎ A penguin has sex just twice a year.

Can You Get Pregnant If You Have Sex Standing Up?

At my school, we were offered a choice between studying Latin and biology. I chose the former and thus I know absolutely NOTHING about the human body and how it works. Oh, sure, I know that we have 206 bones and that the smallest one is in the ear but that comes under the heading of trivia rather than anything more useful.

In fact, so ignorant am I on the subject that I was in my thirties (and already the father of two children) when I first discovered that women had eggs inside them.

How very curious!

I used to believe—until not that long ago—that the gender of a baby was determined by which of the two participants in the act of procreation had tried harder. Honestly.

As I write these words, I'm aware of a lack of knowledge that's almost scary in its implications. I mean, are you familiar with something called "tyrannical ignorance"? It's where you don't know what it is you don't know. In other words, you not only don't know the answer, you don't even know the question to ask.

With so many of the questions in this book, I find myself beset with tyrannical ignorance—and no more so than here.

How does the woman's reproductive system work? Are there bits inside women that might make a difference to the answer?

Who knows? My knowledge of the female form is almost entirely derived from smutty magazines. I feel disoriented when I see a naked woman without staples in her midriff.

Now for the question. By the way, this is what's known in the journalism biz as a dropped intro. Just thought you'd like to know (you can't say you don't learn things here). I always half-believed that having sex standing up was as inconsequential as eating standing up, but that's obviously a case of the wish being father to the thought. So, let's take two couples: Couple 1 and Couple 2. Couple 1 always have sex standing up; Couple 2 always do it lying down. Surely, Couple 1 would be less likely to conceive (always assuming that everything was equally OK with both couples' plumbing)?

I decided to phone the British Pregnancy Advisory Service, whose spokeswoman, Vishnee Sauntoo, is doubly qualified, as she's also a biotechnologist. "No, I'm certain that it makes no difference. I used to work for the Human Fertilization and Embryology Authority—the IVF watchdog, so putting that hat on, I would say that it wouldn't make any difference."

With more patience—and kindness—than I deserved, she went on to explain that it was definitely possible to get pregnant if you have sex standing up. So, let that serve as a warning to all you thickos out there who think that a knee trembler militates against the need for proper contraception.

Some of us (now) know better.

At this point, we had an "aha moment," as Vishnee said that the more stressed you were, the less likely you were to conceive.

Hang on, I said, does this mean that the couple who didn't care about conception were more likely to conceive than . . .

She finished off my sentence: ". . . than the couple who were undergoing IVF treatment and experiencing all the stress that goes with it."

Which explains why supposedly infertile couples can't have children until they adopt and then—hey presto—the woman gets pregnant.

Before letting Vishnee go, I asked her to confirm or deny the theory that having sex in a certain position can help determine the gender of a baby, but her response was, "We're not aware of that."

◎ The female New Mexican whiptail lizard can reproduce without any male contact.

Why Don't Men Blow?

OK, so this is a riff on the title question of the previous book, *Why Girls Can't Throw* (they can, they can), but there is a serious question to be answered.

Why are men so reluctant to perform cunnilingus? Is it disinterest in a woman's pleasure (for the many women who find it pleasurable)? Is it just squeamishness?

Or is it—and I'm sorry, but there's no other way I can put this—the smell?

Specifically, the smell of fish.

Look, I'm not trying to gross you out, but there is a problem out there (or, rather, down there).

So, what causes this smell? Incredibly, according to my researches, it's smegma. Whaddya know? It isn't unique to uncut men. Apparently, the same stuff gathers under the clitoris hood, where, if it's not washed away regularly, it develops into a powerful sexual deterrent.

But why smegma? What is it? Is there any point to it?

Incredibly, there is. With men, it moisturizes the glans to keep it soft and smooth; for women, it performs the same function for the clitoris. For both genders, smegma has beneficial antibacterial and antiviral properties which keep all our nether regions healthy.

But is it a fishy smell or am I merely universalizing my own reaction?

I decided to consult my frontiersman friend, Rob (the guy who answered the questions "Why Don't We Eat Squirrels" and "Why Is It Easier to Park a Car Backwards than Forwards" in *WGCT*). Rob is a man of many answers, but he also has a special expertise when it comes to answering this question. No, not because he's a fine and sensitive lover of huge experience (as he would like us to believe) but because he's totally allergic to ALL fish. Just the merest taste of tuna will cause him to upchuck everywhere. I haven't actually seen him do his projectile vomiting stuff, but he assures me that it is a memorable sight.

Anyway, given that Rob is so allergic to fish, is he also allergic to whiffy pussies?

"Yes, sometimes."

So, there you have it. Proof, if proof were needed, that the smell of unwashed binky is, indeed, piscine (it means "relating to fish," not "relating to a French swimming-pool").

Anyway, I trust that explains why some men are reluctant to blow some women. It's for precisely the same reason that some women are reluctant to blow some men: lousy hygiene. There's no point in men and women taking the trouble to dress up for each other if they're not going to wash themselves first. The neglect of intimate parts will, necessarily, militate against intimacy.

It's interesting to note that our fathers' generation had to be persuaded to use underarm deodorant. Thankfully, BO is now a rarity, and perhaps it's time to address ourselves to our nether regions with the same zeal. Although, I should stress that, except in extreme cases, I am opposed to the use of "hygiene products": soap and water will do nicely.

All right, so that's disposed of one important reason why men don't blow—what about the others?

I consulted a few of my (male) friends, and their answers were all pretty similar, which is, I guess, a function of being middle-aged, married, and a friend of mine.

None of them are—or, at least, none of them admitted to being—keen carpet-munchers.

Why? "Because," said one, who demanded anonymity, "it's about my pleasure, not hers."

Another friend told me it was reserved for "high days and holidays and, if truth be told, we both keep score of who did it last."

Yet another reckoned, "It's not worth the effort—besides, she never goes down on me."

So, there you have it: it's a combination of selfishness and laziness.

I think there might be a bit more to many men's disinclination to perform oral sex (though not, you can be sure, when it comes to receiving it). There's also the male psyche to consider: it's about cutting to the chase, getting to the destination. Any foreplay merely delays the main event and, for many men, is more a necessary chore than a pleasure to be enjoyed for its own sake. After men have achieved their own orgasms, they could start cunnilingusing then, but for so many men—yes, including this one, if I'm honest—once we've come, it's game over.

Obviously, I had absolutely no intention of discussing any of this with a real live woman, but unfortunately such a person got to read it and was extremely indignant.

"The only smell here is the whiff of misogyny," she declared.

Eh?

"I've never known a man who didn't like giving head. It's highly arousing for both parties. You're just an old fogey."

I'm only three years older than you.

"Yeah, but somehow we manage to belong to different generations. Every generation thinks they invented sex, but my generation, who became teenagers in the mid-1970s, believe we invented oral sex, and now, thanks to you, I know that's true."

Oops.

"I suppose it's possible that in a case of extremely poor hygiene, or even poor health, a woman would give off a smell and that that smell might be characterized by fishiness. But it's by no means a regular state of affairs. The natural lubricant doesn't smell unless it's been allowed to get stale. OK, look, on the hygiene front I might concede one thing. That's the fixed-head shower."

What?

"A girl can't have a decent wash under a fixed-head shower. If you're not going to have a bath, then you need a detachable shower head. You've got to get that thing off the wall and play it in the right spot. Which, might I add, is not at all a bad way to start the day."

If you say so.

In a British Survey . . .

Four in ten women and men said they rarely or never give oral sex. 58 percent of men said their partner didn't like receiving oral sex; one in five women said they were not comfortable enough with their genitals to receive oral sex; 45 percent of women said they didn't like performing oral sex, while 21 percent said they were uncomfortable with it.

Is Sex a Good Form of Exercise?

Yes.

A really intensive session is more beneficial than a gym workout (and a lot more fun), and even married sex is better than no sex at all—I'm sorry, I meant "no exercise at all."

For example, for the average-size person, fifteen minutes of "average" (you know what I mean) sex—with five minutes of foreplay—uses up about a hundred calories.

Not shabby.

And to that, you can add the calories used up undressing each other: according to a professor from Italy—where else?—undoing a bra with two hands uses up eight calories, while doing it with just one hand uses up a whopping eighteen calories.

And then there's the French kissing, which expends up to six calories per minute—so a ten-minute full and frank exchange of tongues would make up for—ooh—at least a finger of KitKat.

And that's without including the number of calories burned in chasing after/nagging your partner for sex.

But even if they refuse to play, the lack of a partner is no bar to losing weight through sex. According to something called the Young People's Reproductive and Sexual Health and

Rights Organization, an act of masturbation burns up between one hundred and one hundred and fifty calories.

This leads me to conclude that I'm not overweight because I eat too much but because I wank too little. This is easily remedied.

"Sex—the poor man's polo."

(CLIFFORD ODETS)

Things Which Are Better Than Sex

The anticipation before . . .
. . . The cigarette/cup of coffee/cuddle afterwards
Chocolate *(especially if you're supposed to be on a diet)*
A financial windfall
A really juicy celebrity scandal
Discovering that the parking ticket you've just been given bears the wrong car registration number
Finding a toilet when you're really desperate
A new season of *The Sopranos*
The feeling you had when you passed your driving test
Getting a pay raise
Seeing your favorite rock star in concert

◎ A really tongue-twisting kissing session exercises thirty-nine different facial muscles and can burn up a hundred and fifty calories—more than a fifteen-minute swim or a hill climb carrying a forty-four-pound rucksack. An ordinary peck uses up just three calories.

What Are the Benefits—If Any—Of Swallowing?

Loads. Huge benefits—no question about it.

In fact, the following message appeared in my e-mail box just the other day:

> *Women who perform the act of fellatio on a regular basis, one to two times a week, may reduce their risk of breast cancer by up to 40 percent, a recent study found.*
>
> *Doctors had never suspected a link between the act of fellatio and breast cancer, but new research being performed is starting to suggest that there could be an important link between the two.*
>
> *In a study of over 15,000 women suspected of having performed regular fellatio over the past ten years, the researchers found that those actually having performed the act regularly, one to two times a week, had a lower occurrence of breast cancer than those who had not. There was no increased risk, however, for those who did not regularly perform.*

It was, of course (and alas), a joke.

And I hope that those women who fell for it took it in the spirit it was intended.

Yeah, right.

But are there any benefits?

Before I go any further, do you know the difference between love, true love, and showing off? The answer is, respectively, spitting, swallowing, and gargling.

But are there any benefits?

It's probably no better—or worse—than the average beverage. Let's say that the average emission is .3 ounces and it's nine-tenths water, which has no calories. So, we're talking about .03 ounces of mainly sugar (fructose, to be precise), a little fat, and a tiny bit of protein. The average emission contains about fifteen calories.

But are there any benefits?

Well, it won't cure cancer—or even a cold—but think how pathetically grateful the recipient would be, and then you tell me what the benefits are.

To be more precise, semen contains aboutonia, ascorbic acid, blood-group antigens, calcium, chlorine, cholesterol, choline, citric acid, creatine, deoxyribonucleic acid, fructose, glutathione, hyaluronidase, inositol, lactic acid, magnesium, nitrogen, phosphorus, potassium, purine, pyrimidine, pyruvic acid, sodium, sorbitol, spermidine, spermine, urea, uric acid, vitamin B12, and zinc. Some elements are designed to nourish the semen on its travels, while other ingredients are there to neutralize the acidic environment of the vagina—which would otherwise be inhospitable to sperm.

> ◎ Every time Marlene Dietrich met a man she admired (e.g., George Bernard Shaw) for the first time, she used to unzip their trousers and take out their penises.

So All Those Initials— What Do They Stand For?

We're talking about the sort of initials found on the sort of sites that the sort of chaps who put together the sort of books like this are obliged to visit. Sites where "working girls" (i.e., nonworking girls) ply their trade.

Some of them—like "A," "O," and "BJ" (respectively "anal," "oral," and "blow job")—are clear enough, but others are positively mystifying. So, in the interest of elucidation, here are some of the more common (*le mot juste*) acronyms.

AWO: anal without (without, that is, a condom: in other words, total lunacy for both parties. See also BBBF)

BBBF: bareback butt fuck (bareback means without a condom. See also AWO)

BBBJ: bareback blow bob (see also OWO)

BBF: bareback fuck (again, not advised unless you like playing Russian roulette with your immune system. See also SWO)

BBW: big beautiful woman (fat and/or big-breasted. Judging by some of the photos I've seen, the second "b" is frequently misleading. If they want alliteration, what's wrong with "blubbery," "bovine," or "bleedin' ugly"?)

BDSM: bondage, domination (or discipline), submission (or sadism), masochism (and all that sort of thing)

CIM: come in mouth (facility offered at the end of a BBBJ. Sometimes followed by the words "and swallow." Not the sort of thing you get at home)

CP: corporal punishment (or the Conservative Party, which, in the good old days, were one and the same thing)

CWM: cyber whoremonger (a punter who uses the Internet. Not to be confused with the Welsh word "*cwm*" meaning "village." Nor indeed with the Council for World Mission, which is a Christian charity organization)

DP: double penetration (two men, one woman—you get the picture. More often used in pornography than in prostitution)

OTK: over the knee (spanking. Also known as The English Disease—but not in England)

OWO: oral without (see also BBBJ)

WS: water sports (but rarely performed with water. In the interests of accuracy—and to preempt embarrassing confusion at the beach—should perhaps be renamed "wee-wee sports")

Is It Really Inadvisable for a Sportsman to Have Sex Before a Match?

My idea of a sporting contest being a light game of pool at my local, I'm hardly the man to answer this question.

It's true that in my rugby-playing days, I would occasionally strike it lucky on a Friday night without it notably affecting my performance on a Saturday afternoon, but I was always such a lackluster player that how could anyone have told?

So, I phoned Dr. Roland Powell, my pet MD, but he told me that the jury was out on this one. "Some people abstain because they think it focuses their mind on the following day's contest, whereas others believe that the release of tension through sex helps them. Some experts believe the presence of testosterone in the body can be advantageous to sportsmen in power sports—like weight lifting or sprinting—but whether this is better achieved by activity or abstinence is a matter of debate."

I did some investigating into specific cases and came to the conclusion that much of it is in the mind—either way—and that so long as a sportsman is in the right frame of mind (however he achieves that), then he'll perform to his full potential.

Are Any Foods Really Aphrodisiacs?

You won't be surprised to learn that I belong to the school of thought that says if you're in the mood, then almost nothing will stop you, and that if you're not, then almost nothing will work (except maybe Rohypnol).

Still, in the interests of scientific research, I unearthed the following so-called aphrodisiacs:

Oysters. These are high in calcium, iron, and vitamin A—all of which help with the process of love. In fact, *all* shellfish—especially lobster—have aphrodisiacal qualities. The truth is that if you're taken (or you're taking someone) out for a lobster supper, you're going to be pretty turned on anyway.

Sunflower seeds. Apparently, these have to be "raw."

Cabbage. Now, I like cabbage—especially with roast beef—but I was as amazed to learn about its supposedly invigorating effect on libido as I was to learn about carrots and turnips. Though in the case of turnips, it's only the tops that are of use.

Tomatoes. You always knew that salad was good for you, well, now you can order double portions. And tomato isn't the only thing which will pep up your love life. While you're at it (or, indeed, *before*, if you catch my drift) check out some avocado, cucumber, green pepper, watercress, and lettuce.

Lemons. Maybe this explains why cocktail bars always give you a slice of lemon with your drink: they're trying to get you in the mood. In fact, all citrus fruit—including limes and oranges—are good to eat if you want a healthy sex life.

Honey. This simple spread contains something called the "gonadotropic hormone," which apparently helps to stimulate the sex glands.

Garlic. Now this poses an interesting dilemma: you eat garlic to make yourself sexier but, of course, the more you eat, the less fanciable you become. Hmm. The same problem also applies to that other great aphrodisiac, onions.

Radishes. This raises similar problems to garlic and onions, but also the possibility of wind, which is, let's face it, hardly the greatest turn-on.

Betel nut (if you chew it). If you remember *South Pacific*, you'll remember Bloody Mary, the woman who sells her daughter into prostitution, "chewing betel nut."

Strawberries. Tastier than many of the above "remedies," strawberries will apparently put you in the mood for almost anything.

And so the list goes on—culminating in, of course, chocolate, which, according to all the women I've consulted, isn't just an aphrodisiac but a damn fine alternative to sex itself. Not surprising, when you consider that it contains trace amounts of phenylethylamine, the chemical found in the brains of people who are in love.

And then there are all the weird things—like snake blood and powdered rhino horn—but, frankly, it's not worth the bother, is it?

The truth is, for most humans of my gender, almost anything will do—including the word "the," as Derek and Clive (aka Peter Cook and Dudley Moore) demonstrated in *Ad Nauseam* (1978).

Apparently, grapefruit scent makes middle-aged women seem six years younger to men (it doesn't work the other way around).

Is Fidelity No Big Deal for Gay Couples?

I ask this question because many of the gay men I have known have been (not to put too fine a point on it) utter sluts. One old friend in particular used to come into my office at the BBC in the morning just to tell me how many men he'd had sex with the night before.

I asked him why he was so promiscuous, and he told me that of the two things—coming out as gay and being promiscuous—the former had been far more of a stigma than the latter could ever be.

Mind you, that was some twenty-five years ago, and he had been married with children at the point where he came out.

What about today's gays?

Another gay friend explained gay promiscuity thusly: "With a man and a woman, you have—or at least traditionally you had—the woman saying, 'No, we shouldn't!' With gays, you have two men both saying, 'Yeah!'"

In other words, it's the irresistible force meeting the *moveable* object.

So, does this explain gay infidelity?

Once again, I turned—alas, for him, only metaphorically—

to Simon Fanshawe. First of all, he invited me to consider all the gay relationships where fidelity *is* a big deal. He is, of course, absolutely right: I can think of at least three gay couples I know, who have been together for years and are the model of fidelity.

Still, did Simon know what I was driving at?

"Yes, I do. Relationships in the absence of children have a different dynamic. In relationships with children, there is a kind of construct, a template that they're playing out. Where there are no children, you make different kinds of choices."

So, he doesn't distinguish between gay and straight but between couples with and without children: it makes sense to me.

"I know many people who are each other's life partners. However, they might have one-off casual relationships or more serious relationships that are still clearly secondary to their principal relationship. It was different in the past. There was a time when gay men saw themselves as endlessly promiscuous and lived their lives accordingly."

Like children who have suddenly been given the keys to the candy store?

"Absolutely, a blade of grass was privacy enough for us. Now, we've become a much more accepted part of wider society, we find ourselves behaving in the way that other, straight members of that society behave. In other words, once you set a place at the table, you eat differently with other people than you did when you were eating on your own in the kitchen."

What's the Most Exotic Love Offering Ever Made?

You know that bunch of roses you bought last Valentine's Day?

No, that's not it.

What about the box of candy you got in the gas station when you remembered her birthday at the last minute?

No, nor that.

Well, how about the frilly crotchless panties you plucked up the courage to buy?

No, again.

And it's not just that you're not famous. Even if you cut off your right testicle (and I'd advise against it), it wouldn't begin to compare to some of the truly exotic love offerings in history.

On that basis, neither does Bill Clinton's autographing of a Gap dress for Monica Lewinsky.

No, it's got to be more special, more extraordinary.

So, how about Richard Burton buying Elizabeth Taylor the Krupp diamond and various other unbelievably expensive trinkets (as well as marrying her twice)?

Well, that's close, but, as Monica would vouchsafe, no cigar.

Ultimately, we're not talking about the sort of present that money can buy. So, let me, in reverse order, give you the top exotic love offerings:

In third place: Emperor Shah Jahan, who built the Taj Mahal in memory of his favorite wife.

In second position is Cleopatra, who once dropped two extremely valuable pearls into a glass of wine and drank a toast to Marc Antony; he gave her Cyprus.

But the number one has to be Herod, who gave Salome the head of John the Baptist—just because she asked for it. Now how romantic was that?

Even though the candy would have been a lot more enjoyable.

How Old Is Too Old (And When Old Folk Meet Each Other in an Old Folks Home and Decide to Get Married, Do They Consummate the Marriage)?

Sometimes this research/writing malarkey is harder than it looks. I mean, you'd probably think that all I'd have to do to get an answer to this question would be to pop down to the local old folks' home and ask a few coffin-dodgers whether they still . . .

WHAT????????????

Is it too late to change careers—or, at the very least, book titles? Something less scary, like *How to Wrestle Man-eating Crocodiles*? Gritting my teeth, I phoned the National Care Association (in Britain), where, no doubt through equally gritted teeth, they told me it was "down to the individuals."

Well, thanks a million, guys.

So, how old is "past it"?

I am, I suppose, obliged to say "never." But that's not *strictly* true, is it? I mean, "past it" is when you can't or won't do it any longer or, more likely, when you can't find someone to do it with.

As a man approaching fifty, I have a sort of interest in the subject.

I plucked up the courage to ask a friend from the bridge club—a septuagenarian herself—who does a lot of volunteer work with the elderly. To my relief, she wasn't shocked but giggled and told me that she very much doubted whether any of "her" old people were still "enjoying sexual relations."

I told her I hadn't asked whether they were "enjoying" them but whether they were still having them.

She giggled again and shook her head. "There are probably a few people, but I think by the time you get into your late seventies or eighties, any intimacy is restricted to cuddles—that sort of thing."

People Who Were Still Sexually Active in Their Seventies and Beyond

H. G. Wells
Mao Tse-tung
Bertrand Russell
Victor Hugo
Brigham Young
Benjamin Franklin
Leopold Stokowski
Havelock Ellis (*who, to be fair, was already sixty years old the first time he was able to achieve vaginal penetration*)
Johann von Goethe (*who proposed to a nineteen-year-old when he was in his seventies*)
Colette
Leo Tolstoy

Sarah Bernhardt
W. Somerset Maugham
Franz Liszt
King Louis XIV

Men Who Fathered Children after the Age of Sixty*

Cary Grant (62)
Clint Eastwood (63)
Marlon Brando (65)
Harvey Keitel (65)
James Brown (67)
Yves Montand (67)
Pablo Picasso (68)
Francisco de Goya (68)
Charlie Chaplin (73)
James Doohan (80)
Anthony Quinn (81)
Saul Bellow (84)

◎ Australian miner Les Colley (1898–1998) became the world's oldest father in 1991, when he was ninety-three years old.

* It should be noted, however, that in every case, their partners were a lot younger.

How Young Is Too Young

I'd say under sixteen—which happens to be the British age of sexual consent. Now, come on, Mitch, are you sure? Girls mature much quicker than they did in the past, when the age was set: Isn't there an argument for the age to be changed?

All right then, let's make it eighteen.

I'm not joking—and to prove it, I'll stop having a conversation with myself and address the question seriously. Yes, I know that girls'—and boys' too, for that matter—bodies are maturing quicker but that doesn't mean that they're more mature emotionally or intellectually.

In fact, I think there's a very good argument for making the age of sexual consent eighteen—if only to protect younger people from much older predators. The law should work on a common-sense basis. Society can't stop a pair of seventeen-year-olds—of whatever sexual persuasion—from getting it on, nor should it try to. Even a pair of fifteen-year-olds shouldn't necessarily be stopped from doing what comes naturally (although a little sex education and a lot more parental responsibility wouldn't go amiss).

However, a particularly sheltered, innocent, immature seventeen-year-old might need protection from a particularly rapacious forty-five-year-old (note that I haven't specified gender or sexuality), and that is where the law could and, in my opinion, should operate.

Look, I'm not saying that people only become attractive *after* some arbitrary age of consent: I'm saying that they need protecting until that age. In an era where young girls—aided and abetted by parents who should know better—flaunt their precocious sexuality, it can be confusing for chaps (especially when women in their twenties and thirties are simultaneously posing as young girls). Girls like that might very well be sexy—at least superficially—and there's nothing wrong in thinking so: but thinking is as far as any responsible person should go.

The point is that the law has to set limits. Rightly or wrongly, eighteen is the age of majority, of adulthood. I think it makes sense to extend that age to all acts (aye, including smoking and buying lottery tickets) of an adult nature, and what could be more adult than having sex?

Having said all that, I'm aware that I might be merely reflecting contemporary values. Consider these historical characters, who all had sex with (what today would be) underage girls: Paul Gauguin (who married—at different times—*three* fourteen-year-old girls), Sir Charles Chaplin (whose catchphrase was "the younger the better"), Samuel Pepys, John Barrymore, Howard Hughes, the Marquis de Sade, Errol Flynn, and Fyodor Dostoevsky. Consider also Lewis Carroll (who didn't actually do anything to little girls but seemed to have an inordinate interest in them), John Ruskin (who became infatuated with a nine-year-old girl when he was forty), and Émile Zola (who, in his middle age, fantasized about prepubescent girls).

◎ The world's youngest parents were eight and nine and lived in China in 1910.

The Age at Which Famous People Lost Their Virginity

Age 4

River Phoenix (*his family were members of the Children of God cult which encouraged adults and children to experiment with "God's gift of sex"*)

Age 5

Richard Pryor (*seduced by a seven-year-old girl*)

Age 6

Ike Turner (*with a middle-aged next-door neighbor named Miss Boozie*)

Age 8

Joseph Cotten
Sean Connery (*"I was eight but I can't recall with whom"*)
Federico Fellini (*8 1/2—hence the title of the film*)

Age 9

Lord Byron

Age 10

Rossano Brazzi

Age 11

Casanova
Harold Robbins

Age 12

Jimi Hendrix
Billie Holiday (*the first time she had sex voluntarily*)
Don Johnson

Age 13

Johnny Depp
James Caan
Gillian Anderson
Mae West
Jon Bon Jovi

Age 14

Clint Eastwood (*with a "friendly neighbor"*)
David Duchovny
David Chokachi
James Joyce
Cher
David Niven
Bruce Willis (*"I was a fourteen-year-old bellboy at a Holiday Inn and it was the most incredible experience of my life. This really gorgeous chick started coming on to me, so we went down to the laundry room together. She guided me through it and things got kind of hot down there"*)

Natalie Wood
Jerry Hall

Age 15

John Barrymore (*with his stepmother*)
Shelley Winters
Sir Michael Caine
Peter O'Toole
Art Buchwald
Burt Reynolds
Tina Turner
Jack London (*with a girl who "came with" a boat he bought*)
Charlie Sheen (*with a prostitute. "The problem was she wanted $400 so we used my dad's credit card"*)
Madonna (*in the back of a Cadillac with a guy called Russell*)
Sting
Sally Field

Age 16

Benito Mussolini
Jean Harlow
Brigitte Bardot
Carmen Electra (*"It was in Cincinnati in the backseat of a car. It was not very glamorous and I don't remember it being such a great experience"*)
Sir Richard Branson
Richard Harris
Teri Hatcher
Raquel Welch
Shirley MacLaine

Ursula Andress
Shelley Duvall
Groucho Marx
Barbara Hutton
Jayne Mansfield
Mike Tyson

Age 17

Ginger Rogers
John F. Kennedy
Steven Spielberg
Cyndi Lauper
Joan Collins (*"I was seventeen and he was thirty-three. It was just like my mother said—the pits"*)
Erica Jong
Carrie Fisher
Dyan Cannon
Alicia Silverstone
Donna D'Errico (*in a car*)
Dr. Ruth Westheimer

Age 18

Charles Baudelaire (*on which occasion he contracted the venereal disease that would kill him twenty-seven years later*)
Walt Disney (*on his birthday*)
Barbra Streisand
Anthony Edwards
Brad Pitt
Jamie Lee Curtis
Napoleon Bonaparte

Brooke Shields
Victoria Principal
Vivien Leigh
Leonardo DiCaprio

Age 19

Lillian Hellman
Marlon Brando (*with an older, Colombian woman*)

Age 20

Mira Sorvino
Victor Hugo

Age 21

Marie-Antoinette

Age 22

Dudley Moore
Edvard Munch
Chris Martin

Age 23

Debbie Reynolds
Elliott Gould (*with Barbra Streisand*)
Catherine the Great
Mariah Carey
Hugh Hefner

Age 25

Isadora Duncan

Age 26

Bette Davis

Age 27

Sir Alfred Hitchcock

Age 29

George Bernard Shaw

Age 31

Lisa Kudrow (*"I'm glad I waited till I was married. I decided my virginity was precious, an honor I was bestowing on a man"*)

Age 34

Mark Twain

Married At. . .

. . . Age 13

Josephine Baker
Loretta Lynn

Mahatma Gandhi
June Havoc

. . . Age 14

Marie-Antoinette
Jerry Lee Lewis
Janet Leigh

. . . Age 15

Aaliyah
Stella Stevens
Mary, Queen of Scots
Eva Bartok
Marthe Bibesco
Annie Oakley
Fanny Brice

. . . Age 16

Dolores Del Rio
Sophie Tucker
Placido Domingo
Marilyn Monroe
Sandra Dee
Tom Jones
Doris Day

☺ The women of the Tiwi tribe in the South Pacific are married at birth.

Is It True That Orthodox Jews Have to Have Sex Through a Sheet?

Although I'm Jewish, I don't actually know any Orthodox Jews—and they would certainly have nothing to do with someone as ungodly as me. The fact is that I'm what is known as a "secular Jew"—although I prefer the expression "Jewish atheist," which I would qualify by saying that Judaism is the only religion to permit such an oxymoron.

However, I'm not only irreligious, I'm actually anti-religion: I've got no time for it whatsoever and I regard fundamentalists—of any religion—to be mad and dangerous in equal measure.

Jewish fundamentalists are—at least outside Israel—more mad than dangerous. You want proof of their madness? OK, the laws of Kashrut (that's kosher to you) demand that meat and milk are never consumed together. Fair enough—as any-one who's had a bellyache from eating a hamburger washed down by a chocolate shake will confirm. There might even be a good dietary argument to be made in favor of not having milk in your coffee *after* finishing your meal. But what about separating the crockery used for milk and meat dishes? And what about—and this is where madness really does creep in—having two dishwashers: one for milk crockery and the other for meat crockery?

But then, of course, it has absolutely nothing whatsoever to do with whatever the original rule was designed to address, and everything to do with blind obedience.

So, do they have sex through sheets (so that the man won't come into direct contact with the woman)? It wouldn't surprise me. After all, I do know for a fact that an Orthodox Jewish man is obliged to forgo any physical (even nonsexual) contact with his wife while she's having her period, so it might even be logical for him to whip out a sheet when the mood takes him.

I said at the beginning that I didn't know the answer to this question, but the truth is I'd always believed it to be the case. Sadly, it turns out to be wishful thinking—even though I am, by all accounts, not the only person to labor under this misapprehension.

Rabbi Jonathan Romain is a leading figure in Reform (i.e., sensible) Judaism. He's also someone I've known and respected ever since the days when we were teenagers and he was a groovy guy known as "Cat Romain." So, it made sense for me to consult him.

"No, they don't," he said with a chuckle.

"Really?" I insisted plaintively as if he might change his answer—just for me.

"No, it's a myth."

Damn it!

"I tell you what though, they certainly have a lot of sex. It's not uncommon for orthodox couples to have twelve or thirteen children."

But not through a sheet?

"No."

Shame.

"Here's something for you to consider. There is a time of the week—on a Friday night—when Jews are actively encouraged to have sex."

I pricked up my ears—silence in the back there.

Well, that's God and you and me convinced—now all I have to do is persuade Mrs. Symons.

I knew I should have married a Jewish girl—but then she wouldn't mow the lawn or fix the roof.

Why Is a Red Light Used to Represent Prostitution?

This practice goes back to the U.S. in the late nineteenth century, when prostitutes would place a red light in their window to attract potential customers.

But why?

There are two theories—either of which (or both) could be true.

The first goes back to the Old Testament of the Bible (so that's why they call it the oldest profession). When Joshua was trying to capture Jericho, he was helped by Rahab, a Jericho prostitute, who identified her house with a scarlet robe. This was, of course, before the walls came tumbling down.

The second theory is that it comes from the red lanterns carried by railroad workers, which they left outside the brothels when they went inside.

My instinct is that neither of these two explanations is true (even though they happened). The biblical story goes back too far for such a (relatively) modern practice, while the railroad workers' lanterns doesn't make sense, as red would have become the color to indicate that a prostitute *wasn't* available.

The best guess is that one or two "working girls" decided

to "put on the red light" (© Sting) because they'd told clients that that would show they were "on duty" that night, and the practice spread.

And why red? Because it's the universal color of passion and sex. For example, in Japan, a red-light district is "akasen"—literally meaning "red line"—which is, apparently, completely independent of British or American origins.

"Men will pay large sums to whores/For telling them they are not bores."

(W. H. AUDEN)

◎ Prostitution is illegal in Yemen. Prostitutes can be beheaded in public.

Is It True That the Vatican Has the Greatest Collection of Pornography in the World?

When I told friends I was writing this book, I asked for any questions they might have on the subject. This one—from Clive (thanks, mate)—came right out of left field.

It couldn't be true, could it?

And yet, I had just enough suspicion to find out one way or the other.

Alas, it wasn't true. And I so wanted it to be!

However, Clive wasn't entirely stupid for asking, because there is something of an urban myth surrounding the Vatican and pornography.

It turns out that it came from a throwaway remark by the renowned sexologist Alfred Kinsey, who apparently told his students that he had the greatest collection of pornography—second only to the Vatican's.

It was a joke. JOKE. But, as with so many of these things, it spread (helped in recent years by, you've guessed it, the Internet).

By the way, guess who does have the world's largest collection of pornography?

Yes, that's right—it's the Kinsey Institute in Bloomington, Indiana.

Men Who Collected Pornography/Erotica

King Farouk I
Adolf Hitler
J. Edgar Hoover
Samuel Pepys
Ian Fleming

Men Who Wrote Pornography/Erotica

Alexander Dumas Sr. (*used to send his younger lovers dirty poems*)
William Faulkner (*wrote erotic poems*)
Adolf Hitler (*used to draw pornography*)
Mark Twain (*used to write obscene stories and poems*)

In a Threesome, Is There an Etiquette to Deciding Which Way to Turn?

"I believe that sex is a beautiful thing between two people. Between *five*, it's fantastic . . ."

(WOODY ALLEN)

Despite some serious pleading with a girl I used to go out with in my early twenties, I have never known the joys—or otherwise—of three in a bed. Nowadays, as a happily married man, I feel lucky to participate in a two-in-a-bed romp. Give it a few years, and I'll be only too delighted if I can stage a one-in-a-bed tryst. Such is life.

But others are luckier.

The following people all took part in threesomes: George Gershwin, Jack Johnson, Honoré de Balzac, Lytton Strachey (with Dora Carrington and a man), and Janis Joplin (with a man and a woman).

Meanwhile, these extremely fortunate folk all found themselves in a four-(or, in some cases, more)-in-a-bed situation:

Clara Bow, Sir Richard Burton (fittingly, the explorer—not the actor), Lord Byron, Casanova, Catherine the Great, Cleopatra, Alexander Dumas Sr., Errol Flynn, Paul Gauguin, Janis Joplin, Edmund Kean, John F. Kennedy, Guy de Maupassant, Elvis Presley, Grigori Rasputin, Babe Ruth, Marquis de Sade, Dudley Moore, John Barrymore, and Jimi Hendrix.

Alas, all the people cited above are, of course, dead.

It was time to put in a call to Rick the satyr.

"Sure, I've had threesomes," he said with enviable coolness. "Who hasn't?"

Well . . .

"Basically, everyone except you, Mitch."

Hang on!

"No, I'm just breaking your balls! Threesomes can be fun or they can be a pain—just like regular sex. The one thing I would say is that when it works well, there's nothing better than a threesome, but when it's bad, it's the absolute pits. That goes double when you're drunk—which is, of course, how most people end up in threesomes in the first place.

"The best ones are the planned ones and the best combo is two guys and one girl."

You are joking! I didn't take you for a gay boy!

"For a what? Don't be a tosser—or at least no more than you really can't help. Look, moron, a woman can come an unlimited number of times and has three orifices so, like, you do the math."

Maths.

"What?"

It's "maths" with an "s" in this country.

"Sor-ry," he said, not meaning it.

OK, I take onboard what you're saying but surely it's more fun for a bloke to have two girls, eh?

"God, you are so fucking bourgeois!"

I ignored him. So, how do you know which way to turn? Is there an etiquette?

"What does it matter?"

Come on, Rick.

"All right. Look, if it's your girlfriend who's set up the threesome, don't devote all your attention to her friend. That goes double if the friend's cuter than your girlfriend. Make sure you look after both of them—though you look like someone who'd have difficulty looking after just one of them.

"If you're with two girls, DON'T start trying to manipulate them—you're not a fucking film director—and don't start telling them about your fantasies. Well, at least not unless they ask. Above all, DON'T discriminate between the two."

How do you mean?

"I mean, don't go down on one and not on the other—just because you don't fancy her so much."

Thanks, I'll bear it all in mind for . . . for . . . for . . . the afterlife.

- King Ethelred the Unready spent his wedding night in bed with his wife and his mother-in-law.
- 64 percent of women sleep on the left side of the bed.
- The largest orgy took place in 200 B.C. in Rome. There were some seven thousand participants.

Are There Any Countries That Don't Restrict Gays from the Military?

Before I answer the question, let's make it perfectly clear that throughout history, gays—aye, and lesbians too—have served with distinction in the armed forces.

Many gay men were killed in active service in both world wars and in wars and conflicts ever since.

T. E. Lawrence (of Arabia), Alexander the Great, King Richard I (the Lionheart), Julius Caesar, and Pompey were all gay or bisexual—as we understand the terms today—and no one could dispute their fighting prowess.

Although military and naval authorities have always frowned upon homosexuality (and, indeed, being openly gay was usually enough to earn a soldier in most militaries a dishonorable discharge), on the ground—in the field and on ships—traditionally attitudes have been more flexible.

A soldier friend of mine explained: "As a soldier, you necessarily form attachments to and with other soldiers. These can be deep friendships, and the well-being of the army depends on men having two, three, or four friends who will look out for each other. In wars—especially when nonprofessionals join the army—these relationships can sometimes be sexual or quasi-sexual as well. That was certainly the case in World War II, where some couples were openly gay. However, for the most part, it works on the basis of 'don't ask, don't tell.'"

That was the Clinton line, wasn't it?

"Yes. The other line that came from that era was 'soldiers don't want to take showers with men who want to take showers with soldiers,' but, in fact, you'll find that soldiers are, in some ways, more tolerant than the rest of the population. If a bloke's popular and well-respected then the fact that he's gay won't necessarily prejudice them against him.

"As for the navy, they didn't call it 'rum, sodomy, and the lash' for nothing, you know."

Further research shows that Austria, Australia, Belgium, Canada, Denmark, France, Germany, Ireland, Israel, Japan, the Netherlands, Norway, Spain, Sweden, the U.K., the Czech Republic, and Estonia all allow gays and lesbians to serve in the armed forces.

Hungary, Turkey, Greece, and Italy actually ban gays from serving in the armed forces.

The policies of other Western countries—including the U.S.—tend toward "don't ask, don't tell."

Third-world countries—particularly in Africa—are as antagonistic toward homosexuals in the army as they are to homosexuals in civilian life.

Does Sex Cure Acne or Is It Just That Boys Lose Their Spots at the Same Time as They Lose Their Virginity?

I confess that I hadn't heard of this rumor—not even during my own adolescence—but then I suppose I was lucky enough not to suffer the curse of acne (unlike Dustin Hoffman, Sienna Guillory, Jennifer Capriati, Jack Nicholson, Ricky Martin, Mike Myers, and Sir Derek Jacobi, who, in their younger days, were all regular visitors to Zit City).

Mind you, the first lad in my crowd to pop his cherry did, in fact, have the worst acne of all of us—aye, even after he'd "done it" six times (he kept us up to date with his progress and was decent enough to inform us that you weren't a man until you had done it six times—which was handy for him).

Nevertheless—or do I mean nonetheless, and what, indeed, is the difference?—as I've had reason to remark before, what do I know?

Someone, however, who should know is my old pal (well,

more of an acquaintance, if truth be told), Dr. Roland Powell, who, for once, wasn't so forbearing when I asked him whether sex cured acne.

"No," he said somewhat testily (if testily means what I think it means—and you can't be too careful), "having vaginal intercourse does not cure acne."

Oh, OK. Well, that's it then. Perhaps the rest of the book can be completed in a series of yes/no answers. It'd save time but might possibly turn out a bit boring.

So, Doctor, is there a syllogism at work here? Is it that boys lose their spots at roughly the same time as they lose their virginity but the two are otherwise unconnected?

"I know what a syllogism is, thank you very much," he replied, still somewhat testily (if testily . . . etc., etc.). "I suppose so. What is beyond doubt is that acne is a common side effect of puberty—usually on the face . . ."

Isn't nature a bitch?

". . . It's also true that the acne is caused by an increase in testosterone, an androgen. Androgens are also partly responsible for an increase in the oil—or sebum—produced by the sebaceous glands in the skin which, when affected by bacteria, lead to acne. When the hormonal changes of puberty end—so too does acne. Or at least usually. We're talking about the age of sixteen, seventeen, and eighteen."

In other words, the precise time that most people have sex for the first time.

"Exactly. The other connection between the two is that a youth with a face full of spots is unlikely to appeal to a girl, so he might have to wait until his face clears before he can experience sex.

"There's a similar syllogism with girls, who think that going on the Pill will cure their acne. It might—in some small way—but the truth is that their taking the Pill and their spots clearing up coincide simply because of the age they are when both happen."

◎ Chinese crested dogs can get acne.

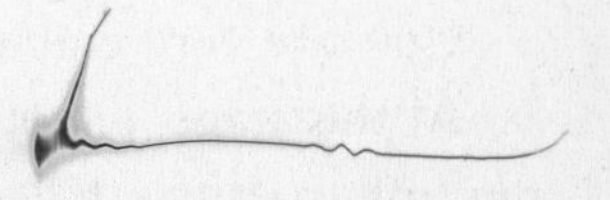

Is It Possible for Men to Fake Orgasm?

Difficult one, this. Unlike a female orgasm, which doesn't (or, at least, not necessarily) produce physical evidence, the male orgasm does.

Trouble is that the answer obliges me to reveal more about myself than I'd rather do. Because how can I, in all conscience, consult an expert when I know the answer to the question is yes.

You know the problem that some men have with premature ejaculation? Well, I used to suffer from the opposite. Sometimes, it would take me forever to achieve a climax and, yes, on some occasions, when I didn't think I would get there, I would pretend that I'd come.

"You haven't come," said one girl when I'd stopped.

"Yes I have."

"That's very strange, I didn't feel anything . . ."

Cue one very embarrassed teenage boy.

At this stage, I should point out that those halcyon days of multiple partners fell post-Pill and pre-AIDS. Consequently, this largely obviated the need for condoms. Taking care of contraception consisted of saying to one's partner, "You are on the Pill, aren't you?"

Anyway, this explains just why girls could tell if their blokes had indeed reached their climaxes.

Obviously, using a condom would have allowed for any amount of subterfuge and deception, but condoms were as alien to us as "unsafe sex" is to my sons' generation. Or so I hope.

Eventually, by the time I reached my early twenties, I realized that I had a problem that was simply caused by a mixture of nerves and—incredible as this will sound to any of my subsequent lovers—an inability to concentrate on my own pleasure. In other words, I would be so pathetically grateful to any girl who'd allow me to do press-ups on her, that I would devote the whole time to giving her a good time. I remember one girl who had eight orgasms to my, er, nought. Hey, now I understand why she used to pester me for years after.

Fortunately—for me, if not for my partners—I learned to achieve an orgasm during sex with others as I did in all my solo encounters. In other words, I could reproduce my training pitch form in the big match.

So, to answer the question—if only from my own necessarily limited experience—I'd say that, yes, men can fake orgasms but not the physical proof . . .

Can a Man Be Raped by a Woman?

There's an old saying to the effect that an erect penis has no conscience—but does it have a choice? That's the nub of this question.

Obviously now, with Viagra, a man can have an erection of which a woman who has managed to subdue him (or, at least, the rest of his body) can take full advantage without his consent.

That would clearly be a case of rape.

Before then, it was much harder for a woman to rape a man and even harder for a man to prove it.

There have been isolated reports—and urban myths—of young men brought to erection by attractive girls and then having a tourniquet wrapped around their penises which would then be used by older, uglier women for intercourse. These "incidents" would invariably take place in factories in backward countries as initiation ceremonies or as punishments for lewd behavior, and should, for the most part, be treated as cautionary tales.

Much more common—and real—are instances where women have drugged, blackmailed, and/or physically threatened men into nonconsensual sex.

However, here the argument against the plausibility of a

woman raping a man always went that even if a woman was threatening that man with a gun, he wouldn't be able to penetrate her unless he achieved an erection—in other words, even if he didn't consent, his penis did.

That strikes me as a specious argument.

Would you say that a woman threatened with a gun isn't being raped if she becomes sexually aroused to the point of lubrication?

Of course not. It's a question of consent.

However extreme seduction might be, there is always a real dividing line between that and coercion.

The (in)famous case of Joyce McKinney illustrates this perfectly.

McKinney was a former Miss Wyoming, who joined the Mormons and began a relationship with a young missionary named Kirk Anderson. He ended it but she continued to pursue him. In 1977, he moved to London, but, with the help of a private detective, McKinney tracked him down and, helped by Keith May, a man who was in turn infatuated with her, kidnapped Anderson and took him to a cottage where he was held for three days.

During that time, McKinney and May tied down Anderson's arms and legs with leather straps, padlocks, rope, and chains, so that Anderson was spread-eagled on a bed. May left the room and then, according to Anderson, McKinney "grabbed my pajamas from just around my neck and tore them from my body. The chains were tight and I could not move. She proceeded to have intercourse. I did not want it to happen. I was very upset."

McKinney countered this by saying, "Kirk has to be tied up to have an orgasm. I cooperated because I loved him and wanted to help him. Sexual bondage turns him on because he doesn't have to feel guilty. The thought of being powerless before a woman seems to excite him. I didn't have to give him oral sex . . . I did do it at his request because he likes it."

Then, in a sentence that kept satirists and comedians in material for the rest of the 1970s, she added, "I loved Kirk so much that I would have skied down Mount Everest in the nude with a carnation up my nose."

McKinney was charged with kidnapping—because it was then impossible for a woman to be charged with raping a man—and given bail to reappear at her trial, which was due to start in May of the following year.

Guess what? She didn't turn up. In her absence, she was eventually sentenced to a year's imprisonment.

But not for rape.

I remember the case well. As a man of the same age as Kirk Anderson, I was mightily amused—and not a little aroused—by the thought of a man being thus coerced by a not-unattractive woman (ridiculously upturned nose, as I recall).

That was then, this is now.

These days, we know better.

Don't we?

Is It True That Eskimo Men Lend Their Wives to Strangers?

Have you ever heard of this one? I had and, I have to say, it was quite a cherished fantasy—until, that is, I actually caught sight of some photos of Eskimo (or, as we must now call them, Inuit) women.

Er, no thanks . . .

They say that people adapt to their conditions. So, while people who live in the Sahara desert are tall and thin, people who live in and around snow tend to be short and squat.

Not my type, I'm afraid.

Still, there was no reason why it might not be true—especially as Eskimos are renowned for their hospitality and, well, it is jolly cold in the Arctic. Hence the name "Arctic."

Once again, I turned to my favorite—not to say my *only*—social anthropologist, Dr. Lorraine Mackintosh.

"Sorry to disappoint you, Mitch, but it's not really true."

Not really? That suggests that there's something in it.

"Well, using the word 'Eskimo' as a broad term for all the people who live in the extreme north, it is true to say that, traditionally, they had slightly different marital arrangements from other contemporary groups. For example, they went in for what might be called 'spousal exchange' . . ."

Wife-swapping?

"Sort of, but it was usually tied in with religious ceremonies. And sometimes these arrangements went much further and developed into a sort of co-marriage, in which families were joined together."

So, there was no wife-lending to strangers?

"Only rarely and certainly not nowadays. I believe that a Danish explorer named Mikkelsen wrote of being offered an Eskimo's wife, but this was probably because of the cold. Nevertheless, his account of the offer fed the legend—along with two other things: Eskimos' unorthodox—to us—marital arrangements and the fact that Eskimo men were renowned for their hospitality."

Why Are Gay Men Called "Fags"?

The British and the Americans are, of course, two peoples separated by a common language.

Nowhere is this more evident than when it comes to the word "fag."

You see, to us Brits, the word means either a cigarette or a young schoolboy who performs menial duties for an older boy—you can see a possible derivation already. Also, the word "faggot" is used to describe a British meatball served in gravy.

Nowadays, of course, fagging (where schoolboys serve other schoolboys) no longer takes place in schools—and, indeed, the consumption of fags (in the sense of cigarettes) is also down. And who eats faggots anymore (who, indeed)?

But fags, as a rude word used to describe gay men, what's the origin of that?

I decided to consult Dr. Caron Landy (who is, incidentally, halfway to being a fag hag: yup, she's a hag).

"If you go back to the origins of the word 'faggot,' you'll find that it refers to a bundle of sticks or branches."

OK, but how do you get from there to gays?

"Patience, patience, you have none. It was thought that there was a connection, because homosexual men were burned at the stake, and the fire would have been started with a bundle of sticks. However, this is unlikely, since this was a British practice, while the use of the word 'faggot' or 'fag' to mean a homosexual man is American in origin. But if we can stay on this side of the Atlantic for a moment, there are other clues to the derivation of this usage. Butchers would throw the 'faggot-end of meat' into the gutters; consequently, street prostitutes became known as 'faggots.'"

Female prostitutes?

"Yes. Also in Britain—in northern England—the word 'faggot' was used to describe a stupid person. So, it's possible to see the beginnings of a derivation in . . ."

Is it?

"Look, not all definitions and origins are clear-cut—especially when it comes to slang. Sometimes, there are many separate roots that come together to produce a word in common currency. The word 'faggot' is a case in point. It has a certain onomatopoeic quality, don't you think?"

I suppose so.

"'Faggot' would have been just one in a series of words used to describe homosexuals at a time—America in the 1920s—when the practice was still illegal, but more and more men were displaying their sexuality. Other such words were 'queer' and 'fairy' and 'queen' and, interestingly, gay men would use them themselves to describe different types of gays—much in the same way as they would nowadays describe one another as 'butch' or 'straight-looking' or whatever."

So, that's it then? It just evolved into its pejorative meaning from a series of different routes?

"Pretty much so. I'm sorry I can't be more specific."

You're sorry . . . Where does fag as in the sense of cigarette fit in? Is there a connection with "faggot"?

"No, the leftover bit of a piece of cloth was known as a 'fag-end.' From there, you have the leftover bit of a cigarette being called a fag-end and then the word 'fag'—not unreasonably—became used as a slang word for cigarette."

And fag—as in one schoolboy looking after another—surely that has a link to gay practices?

"You'd have thought so but no, it comes from the notion that the person performing the task became fagged out—or tired—as a result of having to work hard."

Well, I'm definitely fagged out with this question.

Does the G-Spot Really Exist?

You won't be surprised to know that this isn't really my territory. Although I'm old enough to remember the excitement of the "discovery" of the G-spot, I had quite enough trouble remembering the clitoris—let alone engaging with a whole new erogenous zone.

Yup, that's the kind of lover I am.

But after all the fuss of the 1980s, did it even exist at all?

For the answer to that, we have to go back to the eponymous Dr. Ernest Grafenberg, a gynecologist who, in 1950, wrote an article about the sensitivity of the anterior vaginal wall.

This is where the G-Spot (named after Dr. G. in a 1982 book entitled *The G-Spot and Other Recent Discoveries About Human Sexuality*) is supposedly located.

To be specific, it is said to be a sexually sensitive small area on the upper wall of the vagina approximately two inches from the opening and close to the bladder and urethra.

The "discovery" of this spot led to a sort of vaginal Gold Rush as the women of the Western world shucked off their shoulder pads and—sometimes with the help of their menfolk—dedicated themselves to the task of finding this magic button.

Some were successful but many were not, and struggled to see what the fuss was all about.

Scientists were equally divided. Some said it was the key to female ejaculation and proved that Freud had been right all along about his distinction between the vaginal and the clitoral orgasm; others disagreed and pointed to an absence of necessary nerve endings.

Interestingly, although Grafenberg himself made orgasmic claims for the importance of the area that would eventually bear his name, he also said, "There is no spot in the female body from which sexual desire could not be aroused. Innumerable erotogenic spots are distributed all over the body, from where sexual satisfaction can be elicited; these are so many that we can almost say that there is no part of the female body which does not give sexual response: the partner has only to find the erotogenic zones."

So, there you have it. For those women who've found their G-spot, it definitely exists; for those who haven't, it doesn't.

Do Gynecologists Make Better Lovers?

Now I come to answer this question, I've just realized that there's an inherent assumption in it that all gynecologists are men.

In fact, six out of seven gynecologists are men—which makes it pretty overwhelmingly male, as professions go—but not entirely so.

Anyway, that quibble aside, let's assume that the question refers to male gynecologists.

I don't know any gynecologists, nor do I know anyone who's married to one. Fortunately, my wife is acquainted with two women who are or were married to gynecologists. One is married to a consultant gynecologist, who, she says, is brilliant when it comes to diagnosing things—and that was especially true during her three pregnancies—but that he was a tad unsympathetic with her routine gynecological problems. Is he a good lover, asked my wife? Apparently, yes he is.

The other woman is now divorced from a gynecologist and told my wife that he was a lousy lover—always had been, even when they were first together.

So, what can we deduce from this "evidence"? That gynecologists are better or worse lovers?

I don't think so.

No, what we can conclude is that a happily married woman says that her husband is a good lover (whether or not he is), and that a happily divorced woman reckons that her ex-husband is a bad lover (whether or not he was).

C'est la vie.

All one can say with any degree of certainty is that, unlike the rest of their gender, male gynecologists will have a better idea of physiology. Whether this makes them better lovers is debatable, but at least it means they know what they're doing (good or bad) and where they're going.

Meanwhile, a recent survey revealed that as many as a third of American gynecologists reckon that their sex life is adversely affected by their work, while a British gynecologist was quoted in the papers as saying, "You see so many vaginas during a normal course of a working day that when we leave work we just want to have a rest from them. I suppose it's a bit like a porn star."

Er, not quite, mate—but we take your point.

Which leaves me with a question to which there isn't an answer: Why does a gynecologist leave the room when his patient is undressing?

◎ In Bahrain, a male gynecologist can only examine a woman's private parts through a mirror.

Why Do Gentlemen Prefer Blondes?

There seems to be little doubt that men—never mind gentlemen—find blondes sexy. If you don't believe me, then ask any blond-by-choice woman (aka "bottle blonde") whether she gets more catcalls when she's blond or when she's not, and she'll tell you in no uncertain terms that blondes have more fun.

Always supposing, that is, that their idea of fun is being whistled at by Neanderthals.

So, why is it?

Could it be the paucity of blondes? After all, only two percent of the human race is genuinely fair.

Perhaps, but paucity in itself is unlikely to be the cause, or else we'd all be going for people with harelips or incredible overbites.

More likely, there's a question of cause and effect at work here. Since it is widely accepted—in the West, at least—that men find blond women sexier than other women, then the moment a woman dyes her hair blond, she is making a statement (consciously or unconsciously) that she wishes to look sexier. And so it goes.

But that still doesn't answer the question of why men prefer blondes. I decided to consult my favorite anthropologist, Dr. Lorraine Mackintosh, who is herself a blonde (though, I suspect, not a natural one).

"I don't think it's the shortage of natural blondes that causes men to find them attractive but the fact that they look so striking," she said. "If you look at other species, you'll see that when creatures of either sex have a choice of mate, they tend to pick one that looks exciting and different: just look at the peacock. Why should humans be any different?

"I also take your point about women choosing to be blond and sending out certain signals. Again, if we return to the animal kingdom, the idea is that a creature is putting out a signal that effectively says, 'Look at me, I'm in such good condition that I can afford to care about how I look.'

"For some men, there's an attraction to the innocence they associate with blondness. After all, many more children are naturally blond than adults: blondness is something that we grow out of."

Not you, surely?

"I was a genuine blonde in my youth, believe me! Almost white I was."

I believe you. What happened? I mean . . .

"I know what you mean! The way it works is that if all sixteen of your parents, grandparents, great-grandparents, etc. were blond, then you will be too, for all your life, but the more of them that weren't blond, the less likely it is that you will remain blond."

And do gentlemen prefer blondes?

"Yes, in societies where they are still naturally rare, but there's every reason to believe that if there were a society which was almost exclusively composed of blondes, then men would fight over brunettes."

So, in the kingdom of the natural blonde, the mouse is queen?

"That's right!"

- Enrico Caruso was a bottom fetishist. In 1906, the legendary singer was arrested for pinching a woman's bottom while walking around the Central Park Zoo. He was convicted and fined and it transpired that he often molested women.
- The Irish writer James Joyce had a peculiar fetish: he used to carry around a pair of doll's knickers in his pocket.
- James Boswell, Dr. Samuel Johnson's biographer, used to have a thing about trees—but he wasn't just a dendrophile, he actually used to shag them.
- The French philosopher Jean-Jacques Rousseau would show off his butt in the hope that some disgusted woman would spank it.
- The Japanese writer Yukio Mishima had a fetish for white gloves.

Is There Anyone in Personal Ad-Land Who Isn't "Attractive" or Doesn't Possess a "GSOH"?

I've never placed a personal ad—honest—and I've only ever answered one, on one occasion, but when I tell you what happened, you'll understand why I was never tempted to repeat the experience.

It was back in 1979, when I was young, free, and single. One day, I saw an ad in the London listings' magazine, *Time Out*, that went something along the lines of: "Beautiful blond damsel in distress seeks knight on white steed to rescue her."

I was curious and bored, and so I sent a reply—saying that I didn't have a white steed but I did have a white Morris Marina car.

A few months later, I got a call from a nice-sounding girl. Of course, in those days, I hadn't yet heard the important advice: "If they sound cute on the phone, add fifty pounds."

So, off I went in my white Morris Marina and, a couple of hours later, arrived at the girl's door.

Oh dear, oh dear, oh dear.

She was not (as I'd assumed from the phone conversation)

the same sort of age as me (twenty-two); she was also most definitely NOT cute. I'd say mid-thirties, seriously overweight, and, not to put too fine a point on it, ugly.

To my credit, I stayed for a few hours before making my excuses—but not before I asked her (as discreetly as I could) about the wording of her ad.

"Oh, that," she says, "that was a Christmas present from my friends."

"Ah," says I, "so it was they who described you as a 'beautiful blond damsel'?"

"Yes, wasn't that nice of them?"

I smiled through gritted teeth while calculating (a) how soon I could leave, and (b) whether I had an action against her bloody friends under the Trade Descriptions Act.

So, my experience of lonely-hearts ads is what you might call "jaundiced."

Still, there's clearly a market for them, as you can see from all the newspapers and magazines that run them.

Look, I'm happily married—as we've already established—but that doesn't mean that I don't occasionally peruse these ads (I'm also happy with my car, but that doesn't stop me reading motoring magazines).

And, yes, everyone in personal ad–land really does seem to be attractive and in possession of a Good Sense Of Humor.

Which, of course, begs the question: how come they need to advertise?

Just as I was proposing to delve deeper into this issue, a "funny" arrived in my e-mail inbox, which answers the question of why there seems to be such a gap between what people say of themselves and the (sadder) truth.

Here then is how those personal ad buzzwords translate:

40-ish = 49
Adventurous = slept with everyone
Athletic = no breasts
Average-looking = hound
Beautiful = pathological liar
Emotionally secure = on medication
Feminist = fat
Free spirit = junkie
Friendship first = former slut
New-Age = body hair in the wrong places
Old-fashioned = forget blowjobs
Open-minded = desperate
Outgoing = loud and embarrassing
Professional = bitch
Voluptuous = very fat
Huggable = hugely fat
Wants soul mate = stalker

◎ 35 percent of the people who use personal ads for dating are already married.

Are Transvestites More or Less Likely to Be Gay?

The answer, according to a friend of a friend, who occasionally likes to get into women's clothes (when they're *not* wearing them), is "almost universally heterosexual."

I was astonished, as I'd always kind of assumed that men who wore women's clothes were, in some way, heading past the center point on the straight-gay continuum.

"No," said my friend's friend, who, for reasons that are obvious but which still shame the society in which we live, asked to remain anonymous, "not at all. If anything, there's an argument that says that because we have an outlet for our feminine side, then we're much *more* straight than other heterosexual men."

Blimey!

"I wear dresses and put on makeup because I enjoy the experience. I rarely do so outside the house, because it carries so much stigma. At home with my wife, however, is entirely different and we have a lot of fun dressing me up and making me up, etc."

So, your wife doesn't mind?

"Mind? She loves it! We have our very best sex before, during, and after. You should try it!"

No thanks, it's never appealed.

"Fair enough but you shouldn't knock it till you've tried it."

I'm not being coy here: the idea of getting into a dress (as opposed to getting into a woman's knickers) doesn't do anything for me. However, as with all such things, I'm intrigued to know how someone discovers it for the first time.

"I've been cross-dressing since I was old enough to walk into my mum's room and try on her clothes. I never got off on it or anything, I just enjoyed the experience and the thrill of doing something just a little naughty. When I left home and moved into my own place, I kept a few items of clothing I'd bought from charity shops for dressing up in. When I met my girlfriend—who's now my wife—I told her about it and she was absolutely fine."

What if she hadn't been?

"Then she probably wouldn't have become my wife."

It's that important?

"Yes and no. I don't do it often but it's still an important part of me and I don't think I could have ended up with someone who disapproved of it. I guess if she'd perhaps only tolerated it, it would have been OK but, as it is, it's fabulous."

I think, once again, I'm having to learn the important distinctions between gender, sexuality, and, indeed, role-play.

Certainly, all my further research did nothing to contradict the assertion that the majority of transvestites are heterosexual.

Transvestites

J. Edgar Hoover (*supposedly—though it might have been a smear by his many enemies*)

Cary Grant

Eddie Izzard

Alexander Woollcott
Ed Wood (*as featured in the film of the same name*)
Dan Dailey
Jeff Chandler (*outed by Esther Williams in her 1999 autobiography*)

Can Animals Get Sexually Transmitted Diseases (STDs)?

Yes, according to my pal Lucy, a zoologist (who, as we sip herbal tea, points out to me that the word "zoology" logically ought to be pronounced "zo-ology" rather than "zoo-ology").

"STDs occur throughout the animal kingdom."

Which animals can get them?

"You name them."

You're the zoologist—sorry, zo-ologist—you name them.

"Horses, dogs, koala bears, rabbits . . . even snails. However, they're more prevalent in domestic animals than in animals that live in the wild."

Why?

"Because for STDs to spread, it requires a lot of mating, and there's less mating—less proximity—in the wild."

What's the effect on animals?

"STDs can cause sterility in animals. In doing so, it ends up controlling the population of affected breeds. Is there cause and effect? We just don't know. What we do know is that they can have an immense bearing on those animal groups where a small number of males mate with huge numbers of females. It's called 'harem mating' and it makes it easier for the disease's organisms to spread, because these organisms are actually pretty frail."

Tell me, Lucy, can STDs be passed from humans to animals—or, indeed, the other way around.

She smiled.

No, no, no, I wasn't thinking of myself . . .

"Methinks you do protest too much . . ."

Don't be silly and answer the bloody question, girl.

"No, you're all right . . ."

Not me!

"OK, infected humans who have sex with animals are in no danger of passing on their STD—and vice versa."

How come?

"Because STDs are what we call species-specific. But hey, Mitch, you could always try using a condom . . ."

People Who Suffered from Syphilis (But Not Contracted from Animals)

Oscar Wilde—from a prostitute while at university

Niccolo Paganini—reduced him to a walking corpse with halitosis and teeth held in with a string

Al Capone—caught it in a Brooklyn brothel, and it completely changed his personality

Bram Stoker—led to the tertiary stage of syphilis, general paralysis of the insane

Howard Hughes—caught it as a young man and it led to his eventual mental and physical decline

Paul Gauguin—it eventually killed him

King Henry VIII—it eventually killed him

Franz Schubert—died from it

Karen Blixen—died from it

Friedrich Nietzsche—died from it

Vincent van Gogh—the mercury used to treat it did nothing to ameliorate his mental illness
Guy de Maupassant—it eventually killed him
Henri de Toulouse-Lautrec—it combined with alcoholic poisoning to kill him
Alexander Dumas Sr.—it eventually killed him
Lola Montez—it made her bald
Benito Mussolini—it probably accounted for his megalomania

What Is the Origin of the C Word?

Despite—no, *because* of—the fact that it's just the most offensive word in the English language (particularly for women), I have to say that the "C–word" is one of my absolute favorite words. It might not be as beautiful as "royalties," but there are times when it's a lot more useful: when was the last time you called someone "a complete royalties"?

Its principal meaning is, of course, "vagina." However, the word's usefulness lies in its two other pejorative meanings—"unpleasant person" and "utter idiot"—as illustrated in the following exchange. "You're the second biggest cunt in the world." "Why only the second biggest?" "Because you're such a cunt."

I was going to ask the redoubtable Dr. Caron Landy for her help with this one, but, you know, sexist old fool that I am, I'm kinda reluctant to discuss this particular word with a woman—even a woman like Dr. L., who's as tough an old dragon as ever breathed fire.

So, I went through my books and then put the word into Google . . . and, my word, you get taken to some extraordinary sites. Why, it was at least three hours before I even started on the etymology of the word.

According to the Oxford English Dictionary (OED), the

earliest use of the word is in references to a street named Gropecuntelane. There were many roads thus named, because that's where prostitutes did their business and were, one assumes, "groped" (they were pretty literal in the Middle Ages).

At that time—and, indeed, up until about the sixteenth century, the word was merely descriptive of a woman's private parts and not at all offensive. It was used by Chaucer in the *Canterbury Tales*, but by the time Shakespeare was writing, it was not a word he felt able to use freely (although there are one or two clever puns in his plays).

The word derives from Old German ("kunton") and first appeared in Middle English as "cunte" (incidentally, there are similar words also used to describe female genitalia in other European languages). Although the word comes to us from the Germans, there's also the Latin word "cunnus" (meaning "vulva") to consider, but there are, apparently, no etymological links between the two.

The word was not included in any English dictionary from 1795 until Webster's defined it in 1961. The OED followed suit in 1972.

For anyone who's looking for a (shall we say?) softer euphemism for the female parts—here are some alternatives:

Bearded clam
Beaver
Binky
Bush
Clown's pocket
Cock pit
Fadge
Flange
Foo foo
Front bottom
Fun tunnel
Fur pie
Gash
Glory hole

Hairy cream pie
Honeypot
Love tunnel
Minge
Mound
Mrs. Sphincter's next door neighbor
Muff
Penis holster
Poon
Poonani
Poontang
Pussy
Quim
Slit
Snatch
Tunnel of love
Twat
Wizard's sleeve

Is It Possible to Have an Orgasm without Being Touched?

Yes it is! Some people (lucky sods) can achieve orgasms without being touched at all—just from having the right kind of emotional, visual, or psychological stimulation. However, this privilege is granted to very few people. According to Kinsey, only three or four men out of the five thousand in his sample had experienced ejaculation by fantasy alone. That works out at between 0.06 and 0.08 percent. Interestingly—given that female orgasms are usually more elusive than male—2 percent of females in the sample had reached orgasm merely by fantasizing about erotic situations (without any tactile stimulation).

Apparently, it's possible to train yourself to have such orgasms—through exercise and by learning certain types of breathing.

What Is the Most Innocuous Song/Film/Book Banned for Being Too Sexy?

In the days when it had a virtual monopoly of all (legal) British radio stations, the BBC could and did ban just about anything on almost any grounds.

While you could (just about) see the reasons for banning songs like "*Je t'aime (moi non plus)*" by Serge Gainsbourg and Jane Birkin (though it was unbelievable that *Top of the Pops* tried to pretend it didn't even exist when it reached the top of the charts), "God Save the Queen" by the Sex Pistols, and George Michael's "I Want Your Sex," the decision not to play other songs was clearly absurd—even at the time.

Chuck Berry's "My Ding-a-Ling" was banned because he was singing about his ding-a-ling (and isn't it a crushing indictment of the British music-buying public that this silly little song should provide a rock-and-roll pioneer with his only U.K. Number One?).

Here are some others, though, which slipped through the net:

Walk on the Wild Side (Lou Reed). There are so many allusions to sex, transvestites, and drugs in this 1973 song about a Holly who "plucked her eyebrows on the way, shaved her legs, and then he was a she" that it's extraordinary that it wasn't banned.

Pictures of Lily (The Who). A song about the old five-finger shuffle, which no one at the BBC picked up on.

Blinded by the Light (Manfred Mann's Earth Band). A 1976 song—written by Bruce Springsteen in 1973—which is also a paean to self-abuse: "In the dumps with the mumps as the adolescent pumps his way into his hat."

(I Can't Get No) Satisfaction (The Rolling Stones). While Mick and the lads had no end of trouble with *Let's Spend the Night Together* (having to change the words to "let's spend some time together"), two years earlier, in 1965, this song, which alluded to a girl's period—"come back maybe next week/'cos you see I'm on a losing streak"—had no problem getting through.

See Emily Play (Pink Floyd). This apparently innocent song reached Number 6 in 1967 and received lots of airplay from DJs obviously unaware that the song was all about a girl "discovering" herself, and so the song is inviting us to indulge in voyeurism. Not the sort of thing the Beeb would have wanted to encourage in the year of "Puppet on a String."

John Wayne Is Big Leggy (Haysi Fantayzee). This 1982 song reached Number 11 without anyone in the BBC apparently realizing that the song is all about . . . well, put it this way, "big leggy" refers to a part of a man that isn't a leg . . .

While on the subject, I have a coda to add. On a school trip to the former Soviet Union, we made a short stop in Brest, the Russian border town where the 1917 armistice was signed. We were escorted to the memorial by some charming local students, who were keen to know all about our Western music. Some of our twenty-strong party taunted them and the Communist system for banning decadent acts like Slade and

Sweet. Revenge was taken when a Russian girl innocently inquired whether we'd heard a song she'd recently bought: "Give Ireland Back to the Irish," a single by Paul McCartney's group Wings. No, we were forced to concede, we hadn't: it was banned in Britain.

What Are the Signs That Your Partner's Having an Affair?

Depends on your partner. Like good poker players, successful adulterers don't provide any tells. Maybe that's why I don't stray, because I'd give it away to my wife. She'd know—immediately.

There are, of course, obvious—not to say clichéd—signs.

He/she's paying too much attention/too little attention; buying more presents/fewer presents; blushing when questioned; proving elusive.

I'd have thought the real telltale sign was people being out of contact—especially these days with mobile phones. (Although mobile phones might help adulterers as, in the old days, they'd have to give hotel phone numbers to their partners—which would, of course, indicate where they were—whereas, nowadays, who knows where someone on a mobile phone might be?)

Try Googling "having an affair" (as I did), and the first thing you get—or, at least, the first thing I got—is Illicitencounters.com.

"Married but want a lover?"

There are even dating services for "attached people," which

presumably attract people who reckon they'll be safer cheating with people who've also got something to lose—rather than run the risk of meeting a single bunny-boiler.

On the same Google results page, there were sponsored links for:

Catch Cheating Lover

Guaranteed way to catch cheaters without a pricey private detective.
www.my-cheating-spouse.com

Caught with pants down

Spot the cheating rat. Set bait and trap. Guaranteed result. $29.95
www.catchspousecheating.com

Your Relationship is Over

Unless you recover from the affair by doing this one thing right.
www.SurviveAnAffair.com

Prove He's Cheating, Now

In as little as 24 hours you will prove, beyond doubt, he's cheating.
www.catchacheat.com

Why an Affair, Why Me?

Survive an Affair Regain Confidence
Key Tactics to Stop an Affair Now
Break-Free-From-the-Affair.com

How to Survive an Affair

Find out exactly what to say and do to heal the wounds after an affair.
SurviveAnAffairNow.com

Locate someone's position

Find people, property and pets on a map via mobile phone technology.
www.locatesomeone.co.uk

Should You Stay or Go?

How To Know When or If You Should Leave This Relationship
http://www.StayorGo.com

Clearly there's a lot of it about. And you know how you can tell? Because of the volume of commercial advertising the simple question attracted.

Partners lie. Spouses lie. The market never lies.

In a British Survey . . .

Among married people, 14 percent of women and 21 percent of men said they have cheated on their partner. Among cohabitants, 11 percent of women and 21 percent of men said they have cheated. Women's top reasons for straying: 44 percent said they were attracted to someone else; 32 percent said they wanted reassurance of their desirability. Men's top reasons: 48 percent wanted more sex; 47 percent wanted more sexual variety.

In the Days When Homosexuality Was Illegal, Were Any Famous Men Convicted of Gay "Crimes"?

Sadly—and ludicrously (in these more enlightened times)—the answer is yes.

Here's a roll call of just some of the victimized.

Sir John Gielgud. In 1953, the great actor was fined £15 after pleading guilty to a charge of "importuning" in a public lavatory. In an attempt to avoid publicity, Gielgud listed his occupation as "clerk."

Leonardo da Vinci. The greatest artist of all time was a lifelong homosexual. At the age of twenty-four, he was arrested for going with a seventeen-year-old male prostitute and was jailed for a short time before his friends could get him released.

Montgomery Clift. Just after becoming famous, the actor was arrested for trying to pick up a male prostitute in New York.

Oscar Wilde. As everyone knows, Wilde started a libel case against the father of his lover, Lord Alfred Douglas, for calling him a sodomite, which he lost. This led to a criminal trial in which he was sentenced to two years hard labor in Reading Gaol, where he wrote his immortal ballad.

Bill Tilden. The great American tennis player was arrested in 1964 and sent to a California "honor farm" (a minimum-security facility) for "contributing to the delinquency of a minor."

So Who Was Saint Valentine?

In Ancient Rome, February 14 (or its equivalent) was deemed to be a holiday to honor Juno, the goddess of women and marriage. The next day was the beginning of the Feast of Lupercalia, during which the names of Roman girls were written down on pieces of paper and put into jars. Roman boys would then pick girls at random, and these couples would be together for the duration of the festival. Inevitably, the way these things go, some of the couples ended up getting married.

I only tell you all this to illustrate the fact that, just like so many other days in the calendar (Easter and Christmas are obvious examples), Saint Valentine's Day has its roots in pre-Christian times.

But only just. For Saint Valentine, a bishop of Terni, was martyred in Rome in the third century AD. His crime?

Well, the Emperor Claudius II—known as Claudius the Cruel—was having great difficulty in persuading young men to join the army. He decided that this was because they didn't want to leave their wives and girlfriends, and so, unilaterally (as cruel emperors are wont to do), he banned marriages and engagements. Valentine, the bishop of Terni, was having none of it and continued to perform marriages in secret.

When he was discovered, he was condemned to be beaten

to death with clubs and to have his head cut off on February 14, 270 (or thereabouts). Not a particularly nice way to go, but he was declared a martyr and then became a saint, so I guess it all turned out all right in the end.

So, now, thanks to him, every February 14 we send each other mushy cards, buy overpriced flowers, and place the most extraordinarily silly personal ads in otherwise sensible newspapers. Or, rather, you might, but Mrs. Symons and I don't. For although we are romantically inclined the rest of the year, there is something about Saint Valentine's Day that brings out whatever-the-February-14-equivalent-of-Scrooge-is in us, and we are resolutely non-lovey-dovey all day long.

As for when this business of cards, etc. started, it's reckoned that it was 1477, when one Margery Brews sent a letter to a man named John Paston, addressed "To my right welbelovyd Voluntyne." This is reckoned to be the oldest known Valentine's card.

Having said that, according to legend, Saint Valentine sent the first "valentine" greeting himself. It appears that, while in prison, Saint Valentine fell in love with a young girl—possibly the daughter of the prison warden—who visited him. Just before his death, he wrote her a letter signed "From your Valentine." So, there you have it. Or not, as the case may be.

Is It True That Nurses Will "Relieve" a Patient Who Can't Oblige Himself?

Like most adolescents, I'd heard this from a boy who's heard it from another boy, etc., etc., and then I met someone who told me that it had actually happened to him!

I believed him—oh, God, I believed him! I believed him night after night until I nearly went blind with belief. Bear in mind that when I was a lad, "all" nurses were women.

And then it dawned on me that this friend was the same friend who claimed that his Jimi Hendrix poster had fallen down on the floor of his room at the precise moment of the night of the month of the year when the great guitarist had died, a few years earlier and I came to the conclusion that, sadly—oh so very sadly—the wish was father to the belief.

So, I was going to dismiss the question altogether as an urban myth, but then my conscience got the better of me and, with no little trepidation, I sent an e-mail to the Royal College of Nursing, the British nurses' professional body.

This is what I wrote:

Hello,

I wonder if you can help me.

My name's Mitchell Symons and I'm an author (check me out on www.amazon.co.uk). I'm currently writing a light-hearted book answering all sorts of questions—many supplied by readers of previous books.

One of these is this: Is it true that nurses will "relieve" a patient who can't oblige himself?

This might seem frivolous or even impertinent—most of the questions are—but that doesn't mean that I don't do my utmost to answer them as authoritatively as possible. Hence this e-mail to you.

My telephone number is . . .

Very best wishes,
Mitchell Symons

I got a call back from a very nice press officer named Colin. "Do you want the short answer?" he asked.

My ears pricked up. Yes, please.

"No."

Is that it?

"Well, could you please add that nurses are highly professional practitioners and this is certainly something that is not within their range of responsibilities."

Curses!

Can Eunuchs Have Sex?

Fascinating question; prosaic answer. Some and some. Though there aren't too many around these days—harems not being as common as they once were—eunuchs have always fallen into two types: those without balls, who could still have sex (even if their chat-up lines might be a bit shrill), and those without any equipment at all, who, er, couldn't.

And that's that.

Why Are Prostitutes Called Hookers?

This is really simple. As everyone knows, prostitutes are called hookers because, during the American Civil War, General Hooker wouldn't allow prostitutes—or camp followers—to accompany his troops. Consequently, as an act of revenge, his men called such women "hookers."

Nice story . . . except that it isn't true.

Having done a little research into the subject, I can tell you that it has absolutely nothing to do with the general, as the word "hooker"—used to mean "prostitute"—appeared in print *before* the Civil War.

Apparently, the origin of the word is that prostitutes used to trap or "hook" their clients—hence "hooker."

Not as much fun as the general story, but much more accurate.

◎ In Siena, Italy, there was once a law forbidding any woman named Mary from working as a prostitute.

Are There Any Animals— Apart from Man— That Have Sex for Recreation Rather Than Procreation?

According to my zoologist chum, Lucy, not many—at least, not many if you apply the strict definition to "recreation": that there must be no chance of procreation and also that both males and females are willing participants.

"In fact," says Lucy, "the only two mammals that fit those criteria are dolphins and bonobos."

Bonobos?

"They're apes."

OK. How come other animals don't go in for sex on a recreational basis?

"Wrong question! You shouldn't be asking me why nearly all animals *don't* have sex recreationally but why humans *do*."

Please explain.

"Recreational sex is a luxury in which few species can indulge. Think about it: other animals are far too busy gathering food to have purely recreational sex and, if they had the time, why would they waste their energy on something so nonproductive?

"There's another factor too: when animals mate, they often have to fight other animals for the right to mate—females as well as males. Now while that's justifiable for them where there's a biological imperative, it wouldn't be worth it just for fun."

But bonobos and dolphins know better, do they?

"They're more evolved and therefore more like us—though why bonobos and not other members of the ape family should go in for recreational sex, I can't tell you."

Some Animals and How Long They're Pregnant

African elephant (2 years)
Rhinoceros (1½ years)
Giraffe (1¼ years)
Porpoise (1 year)
Horse (11 months)
Polar bear (8 months)
Cat (9 weeks)
Dog (9 weeks)
Rabbit (1 month)
Hamster (2 weeks)

"It sounds strange for me to be saying this, but I've come around to the idea that sex really is for procreation."

(ERIC CLAPTON)

"My father told me all about the birds and the bees. The liar—I went steady with a woodpecker till I was twenty-one."

(BOB HOPE)

. . . and Following on from That, Do Any Other Animals Masturbate and Are There Any Animals—Besides Man—That Are Sometimes Gay?

I wasn't letting Lucy go that easily!

Well?

"The answer is yes and no."

I don't want that sort of answer!

"Tough, because it's the only answer you're going to get! Obviously, with masturbation—as opposed to recreational sex—there isn't the same risk of being attacked by a rival animal. However, the other factors I mentioned—like a disinclination to waste time, energy, or, indeed, useful sperm—would militate against such behavior."

What about our old dog Blue, a daft golden retriever that used to hump legs, clothes, furniture—anything? He didn't *stop* wanking.

"Domestic pets—especially male dogs—*are* different:

there's no doubt about that, and a male dog will behave much the same as his male owner . . ."

Steady on there, Lucy.

"It's also true that monkeys will indulge in group masturbation. I think what we see emerging from all this is that where animals have the time and the energy, they'll also have the inclination."

So what about the other part of the question: do other species also have homosexual animals?

"Well, I mentioned the monkeys and the group masturbation and there have been quite a few examples of gay monkeys and claims made for all sort of other animals—even including penguins. I think we have to take many of these cases with a pinch of salt as there is a great tendency to anthropomorphize animal behavior: to make it tie in with our own."

Would that explain why so many of these "gay couples" are found in zoos in American cities with a high gay population?

"I think it would. What we call homosexual behavior might just be the result of distress or simply play or grooming: it's interesting to note what happens when female monkeys are introduced into a so-called male partnership."

What happens?

"The monkeys revert to heterosexual type."

Vice Is Nice but Incest Is Best. Discuss

Don't be silly. This is an old gag (if it can be so distinguished), which masks a truly evil act.

Mind you, while we're on the subject of gags, there's a British saying that "you should try everything once—except incest and morris dancing."

I've never had any inclination to do either. But then you should see my dancing—and, indeed, my sisters and mother (sorry, girls!).

This was obviously a deeply unpleasant topic to research—and one I put off for some time. It was also an area where the Internet—usually so useful—was not a good idea, because, beyond straight definitions like "sexual intercourse between persons too closely related to marry," so much of the material on the subject is pornographic and therefore totally unhelpful.

So, I consulted my vast library of sexual books—accumulated over a number of years for instances just like this—and talked with a doctor, a psychiatrist, and a social anthropologist.

There are obviously different forms of incest—father-daughter is probably the most prevalent, but brother-sister is also relatively common.

Two things we can say for sure: it has very little to do with sexual appetite and everything to do with control, identity, and individual family dynamics—particularly when that family is dysfunctional.

The other thing to know is that it cuts across class. This surprised me, because I had always understood there to be two types—stereotypes—of incest: the working man who comes home from the pub drunk and demands sex from his wife, who refuses, and so he turns to his daughter instead. The other stereotype harks back to the large working-class families of the 1930s, where brothers and sisters found themselves in positions of enforced intimacy because of a lack of bedroom space.

True as both of those (types of) stories might be, they are not the whole picture. The upper-class father is no less likely to abuse his daughter than his working-class counterpart—though maybe he doesn't have to get drunk first. Similarly, brother-sister or mother-son liaisons can occur in any type of family—particularly (in the case of brothers and sisters) in broken families.

The fact that incest is relatively rare in (what we might term) civilized societies is because—for all sorts of important evolutionary reasons—a disinclination to have sex with close relatives is hardwired into our DNA. It is even said that brothers and sisters are programmed to be sexually turned off each other's pheromones.

Nevertheless, it does happen.

So, where does it all start?

Right at the beginning.

The Bible is full of incest. If you're going to be literal about it, if Adam and Eve had three sons, how did the human race continue without at least one of the lads having sex with his Mum?

But even apart from that, Abraham married his half-sister, while his brother married his own niece. Meanwhile, Lot was seduced by his daughters, who plied him with wine in order to "preserve the seed."

Moving swiftly forward through the years, we find Caligula raping one of his sisters.

Then there's Lord Byron, who had an incestuous—and, indeed, adulterous—relationship with his half-sister.

Casanova fell in love with a teenage girl, who, he later discovered, was his daughter by an earlier conquest. Nevertheless, he also once wrote, "I have never been able to understand how a father could tenderly love his charming daughter without having slept with her at least once."

There is also strong evidence that Victor Hugo had a sexual relationship with his daughter.

Adolf Hitler had an affair with his half-sister's daughter, who eventually killed herself.

Clara Bow's father was mentally sick and, mistaking his daughter for his wife, had sex with her when she was sixteen, while Rita Hayworth was also the victim of her father.

All in all, I think even morris dancing is preferable.

Married Their Cousin (Which Is Not Incestuous)

Albert Einstein (*his second marriage*)
Lewis Carroll
Queen Victoria
Charles Darwin
Edgar Allan Poe (*she was thirteen*)
Jerry Lee Lewis (*she was thirteen*)
Saddam Hussein

- In Belize, the punishment for any man who has sex with or marries his aunt is a severe flogging.
- All the pet hamsters in the world are descended from just one female wild golden hamster found with a litter of twelve young in Syria in 1930. The species had been named in 1839, when a single animal was found, also in Syria, but it had not been seen by a scientist for nearly a century. Selective breeding has now produced several color varieties.

Why Have I Never Had a Wet Dream (Not Even As a Teenager)?

Unless it has happened to you too, you can have no idea how bewildering it was for me as a teenager to hear the other lads going on about their wet dreams.

Oh, I would snigger—I was a fine sniggerer—and make all the right noises in those discussions, but I was just pretending.

While every other boy in the class was experiencing nocturnal emissions on a seemingly endless basis, I was waking up bone-dry with spotlessly clean sheets.

Was there something wrong with me, I wondered?

It was only years later—in fact, about five minutes ago, if I'm honest—that it occurred to me that other boys too might have been dissembling. Who knows—maybe we were all making it up?

I decided to go back to the master: Kinsey. According to his findings, 83 percent of men had experienced wet dreams or nocturnal emissions—most frequently in their teens, with a decline by the time they reached their thirties. The figure for women and nocturnal orgasms was 37 percent—which was higher than I'd have thought. Incidentally, 13.11 percent of males (compared to 5 percent of females) had their very first ejaculation in the form of a nocturnal emission (orgasm).

So, it seems that most of them—in fact, 83 percent of them—weren't making it up. I was just one of the 17 percent.

Still, it proves I was right to assume that my teenage years were the most likely to produce a wet dream—which is some (but not much) consolation.

What Is It with Porn Stars and Their Names?

I have actually met a porn star. She was a British woman named Linda, who used her maiden name to make porn films and her married name for her painting (she was a seriously talented artist). Or it might have been the other way around—but you get my drift. I was sent to interview her by the editor of *Penthouse* magazine some twenty years ago. Can't think why I was chosen, as I rarely did (or, indeed, do) interviews—I'm much too interested in my own views—and I've never been especially interested in porn films (as a former TV director, I'm far too aware of the technical exigencies to be able to suspend disbelief).

Still, there I was, and there she was: attractive, intelligent, and very frank. As I recall, our only disagreement arose when she claimed that porn films were no different from any other kind of film. As I recall, she compared her films to westerns—asking me why I was so fascinated with her having sex in her films when I don't make any fuss about John Wayne shooting people in his films. In both instances, she averred, it was only acting.

I wasn't having any of it. "Hang on, when John Wayne

shoots someone, they get up at the end of the take; when you fuck someone, they stay fucked."

She couldn't—or wouldn't—see my point, and we agreed to disagree. I wrote my profile, and she sent me a lovely letter thanking me for writing an insightful piece.

And now, for the life of me, I can't remember her surname. But I do know that it was something terribly ordinary—like Jones or Davies or something like that.

These days, porn stars—of both genders—have the most extraordinary names.

Try these—just as examples:

Abbey Jewels
Nikki Charm
Chocolate Delight
Holly Body
Koffee Kakes
Bubbalicious
Horny Henry
Lisa Lipps
Wendy Whoppers
Abundancy Jones
Brandi Wine
Juicy Fruit
Bunny Bleu
Big Daddy Porn
Gwendy Licious
Tiffany Towers
Hope Rising
Amber Angel
Johnny Rocket
Sharon Swallow
April Showers
Darling Desire
Blonde Ice
Ebony Ayes
Purple Passion
Long Dong Silver
Aphrodite Night
Crystal Sky
September Raines
Kayla Kleavage
Agent 69
Quentin Cream
Autumn Haze
Fantasy Kiss

—and there are thousands more like that out there.

Indeed, there are even ways of "working out" what your

porn name would be (if you decided to pursue a career in that field—or, rather, that seedy apartment). Either use your middle name or your first pet as a first name and the (first) name of the street you grew up on as your surname. This means I would be Paul Green or (much better) Snowy Green.

Silly porn star names—and my apologies to any of the above who were actually thus christened—are the fun part of what many people consider to be a pretty tawdry (if lucrative) business.

If you doubt the tawdriness, then consider the effect that video and digital equipment have had on the market. Now that anyone can participate in porn films, anyone does. This has led to the rise of what might be called gonzo filmmaking, and increased competition. In turn, this has led to much more sexual recklessness.

Put it this way: whereas my generation was more than satisfied with Sven and Hans saying to a couple of girls, "You want to come back to our place and party?" followed by twenty minutes of straight sex and simple cum shots (also known as "money shots"), people today obviously demand more.

Something like "double anal penetration" may be legal, and female porn stars (not necessarily the ones I've named above) might happily consent to being thus penetrated. However, the harm done to their bodies—even and including a retirement spent wearing colostomy bags—renders the whole bloody thing beyond the pale, as far as I'm concerned.

In other words, I regard it as perfectly proper for consenting adults to watch other consenting adults having pleasure, but not to watch them inflicting—and receiving—pain.

For these are not westerns, and when the shooting stops, the pain continues.

Even for those "stars" who don't get physically messed up, the future isn't necessarily that rosy.

As you can imagine, I collected those porn stars' names off the Internet. I clicked on to one—not one I've listed—at random, only to discover that she's currently serving three to five years in a state penitentiary for burglary.

Glamorous it isn't.

. . . and Why Are the Men in Porn Movies So Ugly (When You Figure There's No Shortage of Applicants)?

You know those old-fashioned porn films I referred to in the previous answer? The ones starring a pair of bearded Germans/Dutchmen and a pair of obliging blond girls?

Well, tell me that those blokes weren't ugly, eh? And fat. And lousy—truly lousy—actors, who couldn't even deliver their sole line—"You want to come back to our place and party?"—with anything approaching verisimilitude.

Even their coital grunting and groaning—"you're soooo good"—sounded about as convincing as a politician's promise.

Watching these films on a flickering screen or, later, on a primitive VCR, my friends and I would wonder at the casting: "We're better-looking than them," we would shout, "we could be doing that!"

Maybe these blokes' ugliness was the USP (unique selling point) for such films: if girls would put out to/for them, then just think what they would do for us!

Later, we would rationalize that perhaps the men got their parts (so to speak) because they could perform while others were watching, that they could keep it up again and again, and that they had improbable stocks of semen.

Well, either that or they were friends of the director.

Viagra has changed all that.

Now, any man can stay erect for as long as, er, the part requires. All day every day. Viagra will keep blood flowing to the parts that other remedies can't reach.

No need for the "fluffers," who used to have to get or keep the actors aroused; no more cries of "got wood!" as an actor finally gained/regained his erection.

The only problem is that Viagra doesn't help a man produce any more cum than he would otherwise have, but that's where the miracle of filmmaking comes in: cum shots can be cut in—albeit shot from different angles—and then reused (ad nauseam).

Viagra means that the porn industry no longer has to rely on ugly men who can keep it up but can instead select men on the basis of their looks and the size of their penises.

Why, they could even choose men on the basis of their acting ability (although I wouldn't bet on it).

There are still, I am told, "ordinary Joes" employed in skin flicks, but, nowadays, they're not ugly sods and they don't have to be called Sven or Hans.

Is It True That Women's Hairy Nipples Get Hairier If Women Remove Their Hairs?

Oh God, is there *anything* more disgusting than hairy nipples?

There you are, back at some girl's place. You start smooching and groping and—saints be praised!—she responds (or, at the very least, lets you continue). One thing leads to another and—yes!—there's that fantastic moment when you realize that you're going to have sex with her.

Feverishly, you undress each other while simultaneously taking care not to miss a single erogenous zone when . . . you discover that she has hairy nipples.

Aaaargh!!!

As far as I'm concerned, hair is a no-no on women—apart from on heads (the more the better) and "down there" (a well-trimmed triangle). But then, when it comes to women, I'm with Germaine Greer: "However much body hair a woman has, it is too much" (although I suspect that she wasn't coming at the subject from *quite* the same angle).

When you broach this subject with women, one of their

standard responses is to say, "Ah yes, but if you start removing hairs, they only come back even stronger!"

To which you groan and say (sotto voce), "Small price . . ." but you've lost, and the conversation has to move on.

So, is it true?

According to trichologist Frank Cunningham, it's not. "The plain answer is no. Cutting or shaving hair from nipples or upper lip does not increase growth potential. In both cases, though, I would recommend electrolysis to stop the hairs growing permanently."

I accept that electrolysis is not fun. A fine needle is placed down the hair shaft into the hair follicle and then a mild electric current is passed through to kill the follicle. Each hair has to be killed individually, and because only a limited number of hairs can be killed during any one session, it could take many sessions to cure the problem.

However, if it's going to remove hair from the nipples or the upper lip permanently, then surely it's got to be worth it?

Unless you're the sort of woman who wants to be with the sort of man (or woman) who likes that kind of thing. In which case, you're welcome to each other.

◎ Howard Hughes liked his women hairy and forbade them to shave themselves.

How Do Prostitutes Who Cater to Men Who Want to Dress Up As Babies Keep a Straight Face?

This was tossed into the book as a bit of light relief—an opportunity for me to have a giggle with a prostitute who specializes in this sort of thing.

So, I phoned up my old friend Mistress Mia, who, according to her listing, offers services such as smoking fetishes, boot worship, and verbal humiliation, and I kicked off the conversation by asking her why I would go to her for verbal humiliation when I can get that at home.

Tee hee: so far, so lighthearted.

However, when I moved on to the specific question about men dressing up as babies, she didn't seem to want to join in.

Misinterpreting her reluctance as sympathy for men who like to do that sort of thing, I remarked that it seemed harmless enough.

"No," she said, "you don't understand. I don't do it, because you get men who want to fill their nappies and I don't do hard sports."

And then the penny dropped. I had assumed that when

men dressed up as babies, it was all a bit of a joke—because I found it funny. It was only when I found myself talking to someone who's involved in it all that I realized just how serious the whole thing is—for all parties.

Because I consulted a prostitute instead of a punter on this question, my empathy (and sympathy) is with the former, and I can see—all too clearly—how they would have absolutely no difficulty in keeping a straight face. Perhaps the question should have been, "How do they keep the disgust off their face?"

And before anyone—yes, Rick, especially you—chides me for being buttoned-up or judgmental, can I just say that although I recognize that coprophilia (the use of excrement in sex) obviously does exist and, therefore, there's nothing wrong with consenting adults indulging in it, I am also entitled to exercise my rights, which include the right to find certain practices disgusting and beyond the pale.

Is There Such a Thing As Sex After Marriage?

Oh God, I know, I'm just asking, but it's depressing me even to think about it. Put it this way: after talking to my closest ten (male) friends and canvassing them on when they last "did it" and how often they "do it," it isn't very encouraging.

Now, at this point, I should probably 'fess up to the frequency of my own sex life. However, my wife—and it does take two, you know—is understandably reluctant to share the secrets of our intimacy with you (or, indeed, anyone else).

So, all I will say is that my experience is no different from that of my friends—although my sad male ego obliges me to say that I'm getting it more than most of them.

Which, alas, isn't saying much.

Look, the latest (2005) studies prove what any of us could have told them in the first place: passion doesn't last forever.

According to Dr. Enzo Emanuele, the man who led the study at the University of Pavia, the chemical in our brains that characterizes intoxicating passion wears off after a year. Dr. Emanuele's researchers studied a group of men and women aged 18–31: some were single, some had recently fallen in love, and some were in long-term relationships.

The new lovers had high levels of the protein known as nerve growth factor (yup, this is the one that causes the sweaty palms, the euphoria, and the palpitations). The couples, who had been together for over a year, had the same low levels of NGF as the single people.

Dr. Emanuele's report concluded: "Given the complexity of the sentiment of romantic love and its capacity to exhilarate, arouse, disturb, and influence so profoundly our behavior, further investigations on the neurochemistry and neuroendocrinology of this unique emotional state are warranted."

Of course.

But where does this leave us—and, by us, I mean *us*? I guess it proves that while there can be sex after marriage, that sex is likely to be more compassionate than passionate: reassuring more than all-consuming.

And here's the rub, these are the—in all senses of the word—"best" marriages. The relationships I know of where they have the most (shall we say) vigorous sex are usually the most fraught: the ones living on the edge.

Nope, as a man who's old enough to remember when sex *before* marriage was an issue, I'll gladly settle for the deep comfort of the double-bed over the hurly-burly of the backseat of the car.

In a British Survey . . .

70 percent of those together one year said they had retained their passion, compared with 58 percent of those together two years, 45 percent of those together three to five years, and 34 percent of those together six or more years.

- Every day, lovemaking occurs about 120 million times around the world, resulting in 910,000 pregnancies. Saturday night is the favorite time for most people to have sex, and the most popular time for North Americans to do it is 10:34 p.m. precisely.
- According to a poll of men and women—married and unmarried—the worldwide average for making love is 106 times per year. The French (of course) are top (with 141 "performances"), followed by the Americans (138), Russians (131), Australians, British, and Germans (all 112), South Africans, and Poles (both 109). The Canadians do it less (105) but are ahead of the Mexicans (102), Italians (92), Spaniards (82), Thais (80), and, finally, the Hong Kong Chinese (57).

Unconsummated Marriages

LaToya Jackson and Jack Gordon
Burt Lancaster and June Ernst
Zsa Zsa Gabor and Burhan Belge
Zsa Zsa Gabor and Count Felipe de Alba
Jean Harlow and Paul Bern
John Ruskin and Euphemia Gray (*he was shocked to discover on their wedding night that she had pubic hair. She eventually left him for the artist John Millais and they had eight children*)
Rudolph Valentino and Jean Acker
Rudolph Valentino and Natasha Rambova
King Henry VIII and Anne of Cleves

Peter Tchaikovsky and Antonina Milyukova
Catherine the Great and Peter III
Fanny Brice and Frank White
Judy Garland and Mark Herron
Giuseppe Garibaldi and Giuseppina Raimondi

Note also that it took Marie-Antoinette and King Louis XVI seven years to consummate their marriage.

◎ In Minsk, Belarns, husbands are not allowed to refuse their wives sex after two sex-free weeks have passed.

Married Nine Times

Mike Love
Pancho Villa
Zsa Zsa Gabor

Married Eight Times

Elizabeth Taylor (*twice to the same man*)
Mickey Rooney
Alan Jay Lerner
Artie Shaw

Married Seven Times

Lana Turner
Richard Pryor (*four times to two of the women*)
Martha Raye
Barbara Hutton

Claude Rains
Stan Laurel (*three times to the same woman*)
Jennifer O'Neill
Larry King
Robert Evans

Married Six Times

Sir Rex Harrison
Johnny Weissmuller
Gloria Swanson
Hedy Lamarr
Norman Mailer
King Henry VIII
Harold Robbins
Steve Earle
Jerry Lee Lewis
Tom Mix
David Merrick (*twice to the same woman*)

Married Five Times

David Lean
Ernest Borgnine
George C. Scott (*twice to the same woman*)
George Peppard (*twice to the same woman*)
John Huston
J. Paul Getty
Ginger Rogers
Rue McClanahan
Mamie Van Doren
Victor Mature
Eva Gabor

Judy Garland
Henry Fonda
Jane Wyman (*twice to the same man*)
George Foreman
Rita Hayworth
Ingmar Bergman
Tammy Wynette
Clark Gable
Veronica Lake
Richard Burton (*twice to the same woman*))
Constance Bennett
Jane Powell
Xavier Cugat
Tony Curtis
Danielle Steel
James Cameron
Billy Bob Thornton
Martin Scorsese
Joan Collins

Is There Such a Condition As Nymphomania or Is It Merely a Cherished Male Fantasy?

Maybe the word "or" shouldn't have been included in the question—for whether or not the condition exists, it will always be a cherished male fantasy.

Why? Because the idea of a woman who's always up for it is mightily appealing . . . until, that is, you actually meet such a woman and discover the hard (or soft) way that you can never make her happy.

I've never (knowingly) met any nymphomaniacs, but my friend Rick, the satyr (and they don't come much more satyrical than Rick), has and he tells me that it was "abso-fucking-lutely terrifying. I mean, I thought I had an appetite but she was insatiable!"

The dictionary definition of nymphomania is "a disorder in which a woman exhibits extreme or obsessive desire for sexual stimulation or gratification," but since the condition is a nonmedical one, an equally useful definition might be "any woman who wants more sex than the man/men she's hooked up with."

The key word—and the one that terrifies even satyrs—is "obsessive."

I mean, meaningless, dispassionate sex doesn't bother most men—in fact, if we're honest, we spend our lives searching for just such a thing—but if it comes courtesy of a monomaniacal, bunny-boiling compulsive, then count us out.

The first full-length study of nymphomania, *Nymphomania, or a Dissertation Concerning the Furor Uterinus,* was written by an eighteenth-century French doctor named M. D. T. de Bienville. He was of the opinion that the condition was caused or exacerbated by "dwelling on impure thoughts," eating rich food (especially chocolate), reading novels, and, of course, performing "secret pollutions" (i.e., playing with themselves).

Throughout the nineteenth century, women "suffering" from the condition of nymphomania were treated much in the same way as boys caught masturbating (see chapter "Why Did People Think That Masturbation Made You Blind?").

Perhaps we should leave the last word to Alfred Kinsey, whose definition of a nymphomaniac was "someone who has more sex than you do."

◎ Dustin Hoffman lost his virginity with a girl who thought he was his older brother. "She was a nymphomaniac called Barbara. She was 19 and I was 15½. My brother Ronnie threw a New Year's Eve party. It was over before it began but she thought she was making love to my brother and when she realized it was me, she screamed and ran out naked."

Women Who Had More Than a Thousand Lovers in Their Lifetime

Sarah Bernhardt
Cleopatra (*on one evening alone, she is rumored to have given blow jobs to a hundred Roman soldiers*)
Catherine the Great
Clara Bow
Edith Piaf
Lola Montez
Mae West

◎ The actress Clara Bow had a formidable sexual appetite but it is NOT true that she slept with AN ENTIRE AMERICAN FOOTBALL TEAM ON THE SAME NIGHT! Although Miss Bow included in her list of lovers many members of the USC team of 1927 (the team in question)—so much so that their coach had to ban them from Clara's clutches—there is no evidence that she entertained them all at once.

. . . And, While We're on the Subject, What's the Male Equivalent of Nymphomania?

I've already introduced the concept of the satyr—defined as "a man with strong sexual desires." Well, that's not at all dissimilar and, if you remove the genders, it might even be a matter of conjugation: *I* have a healthy appetite, *you* have strong sexual desires, *she*'s a raving nymphomaniac.

One important difference between satyrs and nymphomaniacs is that the former are (supposedly) in control of their sexuality—but, then again, this distinction probably reflects society's suspicion of female sexuality.

Don Juanism—the desire by a man to have sex very frequently with many different partners—is another word for satyriasis. It's named after the libidinous literary character, Don Juan, who was immortalized by Byron in a poem and Mozart in an opera (*Don Giovanni*).

There is another term—again used mostly for men—and that's "sex addict." In recent years, several Hollywood stars—all men—have been named in the press as suffering from "sex addiction." Some of them are even said to have been treated in clinics.

Excuse me for not buying into this, but I don't think so. The truth is that the sort of person who becomes a star is often larger than life—with appetites to match. In Hollywood, everyone gets screwed and everyone gets laid and it "doesn't count on location." It was ever thus. Nowadays, however, the press doesn't turn a blind eye and so, when male stars get caught in flagrante, they'd better have a good excuse for their wives, or else, under Californian law, they're going to lose half of everything.

How much easier to plead "sex addiction" and rest up for a few weeks in a clinic.

"I'm preoccupied with sex—an area of human behavior that's underexplored, in general. And I like virgin territory."

(JACK NICHOLSON)

"I'm never through with a girl until I've had her three ways."

(JOHN F. KENNEDY)

Men Who Had More Than a Thousand Lovers in Their Lifetime

Grigori Rasputin
King Ibn Saud (*three women a night—except during battles—from age eleven until his death at the age of seventy-two*)
Jimi Hendrix
Henri de Toulouse-Lautrec
Pablo Picasso
Casanova (*though modern revisionists now claim that the figure is under two hundred*)

John Barrymore
Charlie Chaplin
Duke Ellington
Errol Flynn
Paul Gauguin
Howard Hughes
John F. Kennedy
Aristotle Onassis
Elvis Presley
Herman "Babe" Ruth
King Farouk I (*he tried to have sex with some five thousand women, but regular bouts of impotence meant that he wasn't always successful*)

"I've never considered myself addicted to anything, but if I was, sex was it."

(CLINT EASTWOOD)

If a Gay Man Marries a Lesbian, Who Does the Dishes?

Incredibly, I happen to know a gay man who's married to a lesbian but, alas, I don't know him well enough to ask him the question. However, having observed the two of them, I have no doubt that he does the dishes—and dries them and puts them away too.

However, that's merely a reflection on them and their relationship. He's a talented, kind, easygoing sort of bloke, while she's a champion ball-breaker. I can see what she sees in him, but God knows what he gets from her—unless there's a sub/dom thing going on of which I'm not aware.

The world of celebrities isn't of much help either. The gay Hollywood star Robert Taylor married the lesbian (or bisexual) Hollywood star Barbara Stanwyck. So, where does that leave us?

In a rather unsatisfactory place.

Obviously, for the sake of the book—no less than for my *amour propre*—I want to know that gay men are more likely to be pussy-whipped and kowtowed than straight men. While at the same time, it would be pleasing to know that lesbians are tougher on men than straight women are.

Who knows, maybe some gay men and lesbians conform to such outdated stereotypes? But I'm gradually learning not to bet on it.

In real life, there's no reason to believe that gay men are any more helpful around the house than the rest of my gender, and there's no reason to believe that lesbians are any more assertive than their heterosexual sisters.

This is one impertinent question that doesn't produce a pertinent answer.

Unless, that is, it just did.

What Did Women Use Before Sanitary Towels?

Yeah, right? And what do I know of women and periods (besides the fact that they're best avoided during them)? Precisely.

Fortunately, I happen to be married to someone who is not only a woman (no surprises there) but is also a fine writer and researcher. So it's over to Penny Chorlton (her maiden name and the one she still uses professionally) for the answer:

Tampons of some description have been used for thousands of years—Hippocrates mentions them, and the Egyptian peasants used either grass or papyrus for the purpose. Ancient Egyptian laundry lists mention cloth pads, belts, and tampon-like items.

Until the nineteenth century, women feared blocking the flow (in case it caused intense bleeding). These must have been messy times.

Granny rags—cloth pads made from old sheets, pillowcases, or other surplus material—were folded and pinned into underwear and served the purpose for most women. When odor became an issue, the homemade pads could be boiled to clean them.

When they traveled, women either took their cloth pads home to wash or threw them in the fireplace. By the 1890s, there were special portable burners in England specifically for burning menstrual pads.

Then around 1880, German doctors began proposing sanitary-wear. But it was the Americans who patented the earliest menstrual devices in 1854. These consisted of belts with steel springs to hold a pad, which sounds pretty cumbersome and uncomfortable. Mind you, considering the other corsetry women were wearing, the idea wasn't outrageous. However, the products really didn't start gaining in popularity until the 1870s.

In 1890 American women could buy the first disposable pad, called "Lister's Towel" (made by Johnson and Johnson). These had loops at either end and were worn with a belt—and nothing much changed for the next few decades. These pads leaked and were not particularly absorbent. It's amazing it took everyone so long to come up with something better.

In World War I, nurses used large cotton pads to absorb blood from wounds of soldiers and would keep changing these as and when the need arose. The nurses carried back this idea, and the women soon started this practice to help deal with bleeding caused by menstruation.

The first commercial tampons were introduced in the 1920s, but even up to the 1970s it was still considered more healthy to use sanitary pads to catch the flow outside the body rather than using something inside.

The first adhesive pads came in in the 1970s, and then gradually the pads became thinner and yet more absorbent.

Natural sponges, animal hides, rags, and old newspapers are still used in poorer parts of the world today, despite being uncomfortable, unpleasant, and posing a risk of infections.

What Is the Origin of the Word "Camp" To Describe Effeminate Behavior?

Rather than approach the somewhat unapproachable Dr. Caron Landy, I decided to do my own research.

The Oxford English Dictionary (OED) gave 1909 as the first time the word "camp" was used to mean "ostentatious, exaggerated, affected, theatrical; effeminate or homosexual; pertaining to or characteristic of homosexuals."

Later, by extension, it meant "banality, vulgarity, or artificiality when deliberately affected, especially with humorous overtones." However, you can still see the link, because, for all sorts of reasons, homosexuals would almost certainly have had a greater sensitivity to nuance.

Like so many words, it is better defined by example than it is by strict definition. So, *The Producers* is camp, *The Rocky Horror Picture Show* is camp, *Desperate Housewives* is camp. *The Sound of Music* became camp, *Batman* was camp on an adult level—as were The Monkees. Dame Edna Everage is camp, Elvis Presley (by the time he was playing Vegas) was unintentionally camp, and Liberace was extremely camp—all the more so because there were so many straight women of a certain age who really did think that Lee was "waiting for the right woman."

According to the OED, the use of the word "camp" in this context is "etymologically obscure." However, there are clearly etymological (you don't think I'm going to let go of a word like that in a hurry) roots in the French verbs *camper*, meaning "to pose," and *se camper*, meaning "to put oneself in a bold, provocative pose."

These roots cross-fertilized with the use of the word "camp" as sixteenth-century English theatrical slang for a male actor dressed as a woman (which was, of course, the norm in those days) and/or the French word for countryside, *campagne*, because French traveling actors often dressed as women in the countryside.

Clearly, the key elements are drag, irony, a conversational style featuring bitchy retorts and malicious gossip and, perhaps most important, the certainty that you were part of a small group that was finding something extremely funny in something otherwise totally straight.

And since "straight" is the key word in that definition, it isn't hard to see how camp became so linked with gays and gay behavior.

Why Do Some People Enjoy Being Spanked And/Or Tied Up?

"Hit me, hit me," said the masochist. "No," said the sadist.

It's an old 'un but a good 'un.

In fact, according to the self-confessed sadist I consulted, that's the wrong answer. "Hit them, I should say—but not necessarily when and where they're expecting."

How, I wondered, did he manage to persuade women (for my anonymous source is straight) to be his victims?

"It's not what you think. Ultimately, it's about pleasure—for both of us—not lasting injury. I take extraordinary pains. Think about that sentence; really think about it."

And the women come back for more?

"Usually. Put it this way, I'm normally the one who calls it quits."

But then he is, of course, a sadist.

Although sadomasochism—defined as "the derivation of pleasure from the infliction of physical or mental pain either on others or on oneself"—is relatively rare in practice, Kinsey discovered that 22 percent of men and 12 percent of women reported having an erotic response to a sadomasochistic story.

However, in any event, there's a lot more to the world of

BDSM (bondage and discipline, dominance and submission) than sadomasochism.

And we Brits, who are (perhaps unfairly) characterized—even by ourselves—as a nation of spankers and "spankees," are not alone.

According to research, some 11 percent of American men and 17 percent of American women (go figure) have reportedly tried bondage, while between 5 and 10 percent of American adults go in for "S&M" (i.e., light discipline) on an occasional basis.

Meanwhile, as my sadist source confirms, you don't need to be a sadist to enjoy a little hanky-spanky.

So, I asked him, why do people do it?

"On one level, it's just fun—and you shouldn't underestimate the importance of fun in sex but then there are other, more intense elements to it. It's all about control and the yielding of control during moments of heightened arousal. It's simultaneously exciting and intimate."

What's the difference between "S&M" and bondage?

"Not a lot at the light level of both. S&M is, of course, sadism and masochism, but usually, it just involves a bit of smacking and discipline."

But why would a man pay to be beaten? I was beaten at school and absolutely hated it; I'd pay good money NOT to be beaten again.

"That's one reaction: the other is to ensure that when you are beaten again, you're in control. Then there are other men, who pay to beat prostitutes, but there are very few girls who are prepared to experience anything more than very light pain."

You sound disappointed.

"I'm a little more ambitious."

In what way?

"In terms of threshold of pain."

But surely that's torture?

"Not when she's begging you for more! Obviously, I have to make sure before—and during—that she's absolutely fine with it, but ultimately, it's a matter of trust."

And bondage?

"Although it's often used in conjunction with discipline, it doesn't involve any pain per se. Usually, it just means people getting tied up or restrained with handcuffs—that kind of thing. You should try it."

If you say so.

"You are soooo boring!"

"It's been so long since I made love, I can't remember who gets tied up."

(JOAN RIVERS)

Sadistic Men

Arthur Rimbaud (*the French poet was a confirmed sadist*)
Michael Rennie (*the actor used to regularly employ a whip as a sex aid*)
Ian Fleming (*the author liked to whip women*)

What Happens to Prostitutes After They Retire?

Of all the questions in this book, this was the one I found the most fascinating . . . and the most frustrating.

I phoned several (current) prostitutes and went on to three different Internet forums, but was met with a mixture of blankness and hostility.

The blankness was, I concluded, due to three specific reasons:

1. I was undoubtedly an idiot who was winding them up or trying to wangle a free sex call.

2. Even if I wasn't an idiot, why should they talk to me—for free (after all, they are prostitutes)?

3. Prostitute wasn't the career they dreamed about when they were younger. Very few women become prostitutes because their lives are going well. Consequently, even if they wanted to help me, it's highly unlikely that they'd have any idea of a life beyond prostitution.

One girl—in her mid-twenties—told me that she was planning to be a waitress one day, when she'd made enough money from prostitution. Now, I know it's axiomatic that you shouldn't deride other people's dreams just because they're cheap, but it's not much of an ambition, is it? I mean, it's not Julia Roberts in *Pretty Woman*. Nor is it Catherine Deneuve in *Belle de Jour* or Shirley MacLaine in *Irma la Douce*.

The only film—or, at least, the only film of which I'm aware—that showed the grim reality of prostitution was Tony Garnett's 1980 film *Prostitute*, which I'm tempted to call seminal but won't.

Research shows that the (mean) average working life of a prostitute is five years. Since this figure includes women who are (what one might call) semiprofessionals—students, housewives, etc.—whose careers are likely to be short-lived, this suggests that there are some women who spend decades working in what is euphemistically called the oldest profession. Servicing, according to the same research, an average of 868 men per year. However, if you take out the part-timers, this would mean that there are some women "seeing" thousands of men every year.

No wonder they don't want to talk to men phoning up claiming to be writers.

Obviously, the most common—common in the sense of usual—thing that retired prostitutes do is get married. Indeed, there used to be a saying that "prostitutes make good wives," supposedly on the basis that "practice makes perfect."

I'd have thought that the contrary would be true: it must be very hard for them not to have a vestigial contempt for men in general, which would hardly be conducive to a happy marriage.

This would hold true even in those cases—the majority—where the men were unaware of their wives' previous lives. Sometimes, however, they find out the hard way. In my researches, I turned up the case of a man who married a woman who turned tricks on their honeymoon. It turns out she was a hooker and couldn't break the habit. This might be an urban myth but I saw enough variations on the theme to be persuaded that there was some truth to it.

Former Prostitutes

Nell Gwynne
Billie Holiday
Mata Hari
Eva Peron
Janis Joplin (*in her teens, she tried—without much success—to sell herself for $5 a time. Later, when she was stranded away from home, she "turned two tricks" to raise the money to get home*)
Marilyn Monroe (*as a (very) struggling actress in Hollywood, Marilyn would provide "in-car sex" in exchange for restaurant meals*)
Jeanette MacDonald (*worked as an "escort" in New York before becoming famous*)
Andrea Dworkin (*the late feminist worked as a prostitute in the Netherlands in the early 1970s, when she was extremely broke and the punters were clearly—forgive my lack of gallantry—even more desperate than she was*)
Lupe Velez
Lotte Lenya (*was a child prostitute*)

What's So Erotic About Autoerotic Asphyxiation?

In Beckett's *Waiting for Godot*, Estragon says to Vladimir, "What about hanging ourselves?" to which Vladimir replies, "Hmm, it'd give us an erection." Estragon concludes by saying, "Let us hang ourselves immediately!"

The first time many people—certainly in Britain—ever heard of this extraordinary practice was when the British politician Stephen Milligan was discovered dead from the asphyxiation that turned out to be more successful than the autoeroticism.

There were rumors, too, about the deaths of Jerzy Kosinski and Michael Hutchence.

In any event, it's clearly an awful way to check out—if only because it's so incredibly humiliating when it all goes wrong.

And that's the point, of course: we only tend to hear about autoerotic asphyxiation—or "scarfing," as it's also known—in the breach. In other words, when something goes wrong.

So what precisely is it? I asked my friend Rick.

"It's self-strangulation—usually with a ligature or tourniquet to restrict blood flow—while masturbating. Some people—

like Stephen Milligan—combine it with self-bondage. And that's just plain crazy."

Had he himself, I wondered, tried it?

"No! Do I look like a complete ass? Hey, don't answer that question."

So, why do other men do it? And does it differ from the thrill you might get from playing dice with death in any other way? In other words, is it just a case of (*pace* The Eagles) taking it to the limit?

"There's a lot of that to it, clearly. And I think that the process of restricting blood to the brain is in itself meant to heighten sexual pleasure. Me, I prefer a little bit of toot on the tip of . . ."

All right, let's not go there. So, why do people do it?

"Search me. It's mostly adolescent boys—or men who haven't quite grown up. It's terribly dangerous though. Loads of people die every year from it."

How?

"The asphyxia bit. They lose control of their own consciousness and kill themselves unintentionally."

So, it's suicide.

"No, I don't think so. No more than any other extremely dangerous practice—like BASE jumping—is suicidal. It's about risk. Some of the men who do it rig up devices to wake themselves up in case they forget to breathe."

Didn't Stephen Milligan use an orange or something?

"That's right, but, alas, it didn't work . . ."

What a way to go!

"Yes, it reminds me of that Bill Hicks routine, where he says that, in the event of his unexpected death, he's instructed a friend to go round to his place and clear out all the porn—so that his mother doesn't see it."

After talking with Rick, I did some research and discovered that autoerotic asphyxia (or AEA, as it's known) claims the lives of between 250 and 1,000 young American men every year.

The reason why that figure is so vague is because parents of victims are understandably keen to have a different cause of death—anything, damn it—on the death certificate.

Interestingly—and even more tragically—the victims are usually well-adjusted, nondepressed, high achievers. There is also a tendency toward transvestitism and other forms of dressing up/role-play.

Girls, too, try AEA, but much more rarely, and because when they're found dead they tend to be naked with just the ligature around their necks, parents (understandably) assume suicide rather than sexual misadventure.

In 1981, FBI researchers Hazelwood, Blanchard, and Burgess identified six defining characteristics of AEA typical death scenes for use by police investigators:

1. Evidence of asphyxia produced by strangulation either by ligature or hanging, in which the position of the body or presence of protective means such as padding about the neck, indicate that the death was not obviously intended.
2. Evidence of a physical mechanism for obtaining or enhancing sexual arousal and dependent on either a self-rescue mechanism or the victim's judgment to discontinue its effects.
3. Evidence of solo sexual activity.
4. Evidence of sexual fantasy aids, props, or pornography.
5. Evidence of prior dangerous autoerotic practice.
6. No apparent suicide intent.

How Does a Gigolo Get a Hard-On?

There are many gigolos out there—only they don't call themselves that. "Male escort" is the usual euphemism—although many chaps advertise themselves as "male masseurs." According to Jake, the gigolo I ended up consulting, the difference between an escort and a masseur is that the former "usually goes out for dinner with a lady before they have sex."

So it's a bit more upmarket?

"Usually, but not always: I've done both and I'd say the difference is that the ladies like it a bit—how can I put it?—raunchier when you go round to their place, which is usually a hotel room."

At this point, I should introduce Jake. He's in his mid-thirties and is a very convivial chap. I wouldn't say that he's handsome (as such) but the overall effect is a good one. He's clean and tidy and—as he points out to me—doesn't bite his nails. "Unlike you—my God, you bite your nails so badly!"

On the contrary, I reply, I bite my nails extremely well. However, I think the subtlety is lost on him.

But it's certainly true that he's well turned-out in a dark jacket, pinstriped shirt and tie, and chinos. "I have to be, in my job. Even if I'm not working, I never know when I might get a call."

Do you have regulars?

"Oh absolutely! There are five women who I see regularly—perhaps twice a month."

And do you always have sex with them?

"Usually, though there's one client who tends to use me to accompany her to business events—she's in the media or something and she has to go to lots of dinners."

How does she describe you to people?

He thinks about it for a second. "She doesn't really. I'm just Jake. I suppose she lets them think that I'm her boyfriend."

What do you say if anyone asks you what you do for a living?

"I'd just say I was in business—which I suppose, in a way, I am! Funny thing is, people very rarely ask me."

Too busy talking about their own careers?

He smiled in a noncommittal sort of way.

What about other clients?

"Yes, I have sex with them."

And . . . ?

"It's fine."

OK. Does it, er, work every time?

"For them? You bet! I know how to treat a lady like a lady."

Suddenly, I was given an insight into how he can do what he does: you have to possess absolutely no sense of irony—and clearly the sort of man who can say "I know how to treat a lady like a lady" without blushing is an irony-free zone.

Do you use Viagra?

For the first time in our conversation, he looked annoyed. "No I don't. I don't have to."

I'm sorry . . .

"No it's not your fault but people always assume . . ."

Well it must be good to know that you could.

"I suppose so but who wants to walk around with an erection all evening?"

I see.

"I don't need it."

What, never?

"No."

Really?

He smiled and pointed down at his tool (the operative word): "It's got a mind of its own. Also, I really do like my clients. I couldn't do my job properly if I didn't. It's not just about the sex, you see."

Is a Good Wank Better Than a Lousy Blow Job?

I had a lot of fun running this one past the lads, I can tell you.

Nearly all of them started off by saying that any blow job was better than a wank, but as soon as I invited them to be serious about it, they all concluded that a good wank was indeed better than a lousy blow job.

In fact, one or two went so far as to say that a good wank is better than a *good* blow job.

The arguments cited in favor of "a good wank" included:

"You don't have to cuddle yourself afterwards."

"You don't have to say thank you afterwards."

"It's quicker."

"You're not obliged to reciprocate."

One friend said, "Depends who's doing the blow job," which is fair enough.

Another—how can I put this?—more intellectually challenged friend said, "Depends who's doing the wank," which is, of course, idiotic.

All in all, we can draw two conclusions from this scientific study. The first is that a good wank is better than a bad—or even an indifferent—blow job. The second is that my friends are a bunch of onanistic misogynistic Neanderthals.

Well, why else would they be mates with me?

What Is Uncircumcision?

Uncircumcision—or epispasm—is the process of restoring a foreskin. It is done for various reasons: cultural, sexual, psychological.

As most people know, Jewish male babies are circumcised soon after birth, and so a circumcised penis has often served as a means of identification for those people hunting Jews—especially in societies like Nazi Germany where non-Jews were rarely circumcised (and then only for medical reasons).

As a result of the Nazi persecution, many Jews in occupied Europe resorted to extraordinary means to restore their foreskins so that, ultimately, they would pass scrutiny.

Some men were lucky enough to have (albeit rudimentary) surgery that involved skin grafts. Other men used tape to stretch what was left of their foreskins. Unfortunately for them, this process takes years before it produces any real results. Consequently, ever-desperate men were forced to adopt all kinds of subterfuge: the teenage Roman Polanski, for instance, fashioned a dummy foreskin out of candle wax.

Since World War II, there have been other reasons for the practice of foreskin restoration. In the U.S. especially, where circumcision was incredibly popular (though not necessarily with the boys who were having their foreskins sliced off) throughout the 1950s and 1960s, there have been many men who have subsequently seen themselves as victims of medical fashion.

Such men turn to a variety of remedies in their drive to return to what they see as a "whole state."

Nowadays, there are many surgical and nonsurgical means by which foreskins can be effectively replaced.

The current medical thinking is that circumcision is harmless but unnecessary. According to a (Jewish) urologist I consulted, "If people want to circumcise their baby sons for religious or cultural reasons then that's fine, but otherwise there's no reason to."

Aren't circumcised penises cleaner?

"Yes but that's no argument for cutting off foreskins—any more than you'd cut off fingers because they can get dirty. Boys just have to be taught to wash themselves thoroughly. Same's true for men like us: just because we haven't got a prepuce . . ."

"A what?"

"It's the posh word for foreskin. Anyway, just because we haven't got one between us doesn't mean we don't have to keep ourselves clean."

But don't some boys have to be circumcised when they're older because their foreskins are too tight?

"Yes—and your point is?"

Well, isn't it better to circumcise them when they're babies, because it would be less painful?

"By that logic, you'd also whip out their appendix and their tonsils—oh and their spleens and their gall bladders, just in case."

So, if in doubt, best not to circumcise?

"I'd say so—unless there's a religious imperative and, even then, I'd have it done in a hospital."

What about female circumcision?

"Don't go there."

Really?

"Yes, there is absolutely no advantage to it and the only reason why it's practiced is to subdue women."

Clearly you feel strongly about it.

"Yes I do—and so should everyone."

Does Penis Size Vary with Ethnicity?

If I'm honest, this is really just a polite way of asking "Are black men better endowed?"

The question arises because when I told friends that I was doing this book, I asked them for possible questions—and this was one that raised its ugly head (so to speak) more than once.

There's surprisingly little data on the subject, and you'd better believe that I wasn't about to start looking on the Internet (there are some things I'm not prepared to do—even in the name of research).

So, I contacted the world's greatest authority—the Kinsey Institute—to ask this question, and they wrote back to say that, "As for penis size, yes, there are ethnic variations, as there are for most body parts."

Which sounds fair enough.

The original 1948 Kinsey Report found that the average

white man's penis was four inches (non-erect) and 6.2 inches (erect) while the average black man's penis was 4.3 inches (non-erect) and 6.3 inches (erect). Not much of a difference—especially when erect—and it should be pointed out that Kinsey interviewed many more white men than black men.

So, why do so many white people automatically assume that black men are better endowed? I decided to ask social anthropologist Dr. Lorraine Mackintosh.

"Well," she said, "from what you tell me, they're not, so now we have to try to find out the reason for what is clearly a prejudice."

A reverse prejudice, surely?

"A prejudice nonetheless. My guess is that it will have started when Africa was first explored by white people. They would have been fully clothed but the natives wouldn't have been, and so they would have noticed their penises and, no doubt, remembered the bigger ones.

"Then we move to the United States during the days of slavery. Male slaves were treated as objects or animals and were often made to impregnate female slaves to produce new generations of slaves. So, having been cast as sexual creatures—is it any wonder that the stereotype stuck?"

So, clearly it's not a good thing.

"No, it isn't. Look, you're Jewish, right? Now, although to tell someone that they're good with money is a compliment, if I said, 'Hey, Mitch, you Jews are good with money' or made some other allusion to it, I don't think you'd be grateful."

Well, I would, because I'm not particularly good with money as . . .

"You know what I mean. I would be doing it to reinforce a

stereotype that is, in turn, part of a whole series of stereotypes that were once used to victimize Jews."

Just as being well-endowed is part and parcel of the whole racism shtick? All right, I accept that.

Is It Morally Worse for a Man to Rape a Nun Than a Prostitute?

Having discussed this question on more than a few occasions, I've come to the conclusion that it is.

Although feminists, who tend to take an absolute line on this issue, hold that the rape of any or all women is equally repugnant, I think there is a difference.

First of all, I am not seeking to absolve the man who rapes a prostitute: he deserves to face the full weight of the law. However, there is, I believe a different *quality* to the rape of a nun than there is to the rape of a prostitute.

Assuming—purely for the sake of argument—that, in both cases, the physical damage is equal: a rape committed with the threat of violence but without (as is so common in such cases) any additional injuries, one is left with the damage done to the victim.

Without riding roughshod over the feelings of the prostitute, I would argue that, although she didn't consent to having sex with her assailant, the act of sex is not one that is foreign or (presumably) especially abhorrent to her.

For the nun, however, who has taken vows of chastity and celibacy, it might very well be much more harrowing.

By the same argument, it would be less traumatic for a millionaire to have £20 stolen than it would be for a pauper.

However, even as I write this, I am aware that my conclusions are, ultimately, based on instinct. True, I am a wise, intelligent man who wrote an award-winning opinion column for a national newspaper, but the only thing that separates me from the saloon-bar pundit is that I have a platform for my views.

It's time to bring in an expert.

Nicholas Fearn is just the fellow. He's not only a brilliant philosopher—and the author of *Philosophy: The Latest Answers to the Oldest Questions* (thoroughly recommended)—he's also a thoroughly wise owl.

So, Nick, is it morally worse for a man to rape a nun than a prostitute?

"I think it is.

"Stealing a car isn't made any more excusable just because the car happened to be up for sale at the time. On the other hand, it would be worse to steal one that you knew held sentimental value for its owner. So yes, raping a nun would be worse.

"It reminds me of a different question I've been working on that you could phrase like this: Does a rapist commit a less despicable act if he is incredibly handsome and charming than if he is ugly, weighs three hundred pounds, and smells? Possibly, and it would be naive to think that this is not a factor in whether date rapes get reported. The male model has still done a terrible deed, but there is certainly some difference here. Whether we like it or not, the moral character of our actions depends to an extent on who and what we are. Everyone who works in an office knows this, as it is the difference between sexual harassment and flirting. If a homeless person steals a loaf of bread, his action is clearly different from that of a millionaire who steals it for the thrill. We are used to the debate being about whether the consequences of an action are what makes it right or wrong, or whether the immorality is intrinsic to the act regardless of its repercussions. But the agent himself has been forgotten in all this, and for understandable reasons—for it would lead us to deny the principle of equality before the law and accept something that looks a lot like hypocrisy. Namely, that certain behavior is indeed acceptable (or slightly less horrendous) for some and not others. We are getting to a stage in this country where a lot of people would like freedoms to be selective and laws to be 'tailored.' For example, they would like the middle classes to be allowed to take drugs without the chief of police threatening to come and arrest them, but that doesn't mean they want a free-for-all. To deny the same rights to people in the projects (those most likely to be harmed by drugs), or to those who have shown that they cannot handle the drugs' effects without harming others, is libertarian good sense rather than double standards. The law remains a one-size-fits-all institution in a day when there is no such thing as standard behavior. But it's changing, though admittedly there will probably never be any tailoring when it comes to rapists."

That's the trouble with this question, as with so many others in this book—you ask one question and it raises a whole lot more.

Where Do Nudists Keep Their Hankies?

It's a thought, isn't it?

After all, presumably nudists (or naturists, as they call themselves) have to blow their noses from time to time—just like the rest of us.

So, I wrote to Andrew Welch, the commercial manager for the Central Council for British Naturism, and received the following reply:

"Bathing costumes, as far as I am aware, do not have pockets, so, on a beach, where does *anyone* keep their hanky?

"Naturism is a healthy pursuit, naturists being attracted to the outdoor life and benefiting from regular exposure to fresh air and sunshine—the latter little and often, rather than the usual British thing of roasting for two weeks on a once-a-year holiday—and so the incidence of colds is reduced . . . and therefore a reduced need for handkerchiefs, anyway.

"Naturism, in the U.K. especially, is also practiced indoors in the winter months, in indoor pools, saunas, steam rooms, and so on. The nature of this kind of activity is associated with warm places, but obviously, other than when enjoying the various facilities—typically a two- or three-hour session once every two weeks or so—naturists will be dressed the majority of the time in the winter, when colds are more likely to strike.

"Even in the summer, though, naturism is mostly practiced when the weather and circumstances are appropriate, and even in the hottest places, many people will be dressed much of the time—slipping on a T-shirt in the heat of the day, putting a pair of shorts on to visit the on-site shop, dressing to make the journey to the local naturist beach or tourist attractions, and dressing to eat in the on-site restaurant in the evenings when the sun has gone down. In fact, strange though it may sound, most naturist holiday resorts around the world insist on clothes being worn when dining in the evenings.

"I think there is an assumption that naturists are always naked. I'm sure there are some that take the opportunity whenever and wherever they are, but the vast majority will only strip off when, as mentioned, the weather and circumstances are appropriate. They must also live within the norms of twenty-first century Britain and that means being dressed outside the house, club, back garden, beach, etc., and so, for the majority of the time, there are plenty of places for a naturist to have a hanky."

Well, I think that answers the question.

Some Thoughts on Nudity

"There's something therapeutic about nudity . . . Take away the Gucci or Levi's and we're all the same."

(KEVIN BACON)

"Clothes make the man. Naked people have little or no influence on society."

(MARK TWAIN)

"If I took all my clothes off I wouldn't be sexy anymore. I'd just be naked. Sex appeal is about keeping something back."

(JENNIFER LOPEZ)

"I was born naked and I'm going to die naked, so I don't see anything wrong with it."

(JUSTIN TIMBERLAKE)

"Not all the Greek runners in the original Olympics were totally naked. Some wore shoes."

(MARK TWAIN)

"I come from a country where you don't wear clothes most of the year. Nudity is the most natural state. I was born nude and I hope to be buried nude."

(ELLE MACPHERSON)

"Take off all your clothes and walk down the street waving a machete and firing an Uzi, and terrified citizens will phone the police and report: 'There's a naked person outside!'"

(MIKE NICHOLS)

"I can't bear being seen naked. I'm not exactly a tiny woman. When Sophia Loren is naked, this is a lot of nakedness."

(SOPHIA LOREN)

"If God wanted us to be naked, why did he invent sexy lingerie?"

(SHANNEN DOHERTY)

"It is an interesting question how far people would retain their relative rank if they were divested of their clothes."

(HENRY DAVID THOREAU)

◎ Volleyball is the most popular sport at nudist camps.

Naturists

Benjamin Franklin
Moby
Dame Helen Mirren (*named British Naturist of the Year 2004*)
Jack Nicholson (*at home*)
Walt Whitman
Teddy Roosevelt
John Quincy Adams
Galileo
Elle MacPherson
John Lennon

◎ In 1935, the police in Atlantic City, New Jersey, arrested forty-two men on the beach. They were cracking down on topless bathing suits worn by men.

◎ Men are four times more likely to sleep naked than women.

Postscript

So, what have I learned?

Firstly, that there's more to life than I thought.

From there, I think I've learned to accept more and judge less. However, when and where I do judge, I feel more entitled to do so.

More specifically, I've learned the following:

Not to underestimate the appeal of chocolate to women

The difference between a vow and a promise

Missionaries don't necessarily have sex in the missionary position

Women who have sex with other women are lesbians and NOT "lesbian women" (before writing this book I was guilty of that tautology)

It's not a compliment to imply that black men have bigger penises than white men

The meaning of the word "deracinate"

Naturists sometimes keep their clothes on

Acknowledgments

This book couldn't have been written without a lot of help. The key players were (in alphabetical order) Luigi Bonomi, Penny Chorlton, Joy O'Meara, Jeanette Perez, Mari Roberts, David Roth-Ey, Brad Wetherell, and Doug Young.

In addition, I'd also like to thank the following people for their help, contributions, and/or support: Gilly Adams, David Archer, Jeremy Beadle, Marcus Berkmann, Paul Donnelley, Jonathan Fingerhut, Alan Fox, Peter Freedman, Jenny Garrison, Ian Irvine, Brian Johnson, Sam Jones, Martin Lewis, William Mulcahy, Amanda Preston, Nicholas Ridge, Charlie Symons, Jack Symons, Louise Symons, David Thomas, Katrina Whone, and Rob Woolley.

Apart from the Internet and articles in newspapers and magazines, I relied heavily on advice and information from friends and colleagues whom I have credited (see above) or as they appeared in the text.

ALSO BY MITCHELL SYMONS

WHERE DO NUDISTS KEEP THEIR HANKIES?

...and Other Naughty Questions You Always Wanted Answered

ISBN 978-0-06-113407-4 (paperback)

WHY GIRLS CAN'T THROW

...and Other Questions You Always Wanted Answered

ISBN 978-0-06-083518-7 (paperback)

THAT BOOK

...of Perfectly Useless Information

ISBN 978-0-06-73254-7 (paperback)

THIS BOOK

...of More Perfectly Useless Information

ISBN 978-0-06-082824-0 (paperback)

THE OTHER BOOK

...of the Most Perfectly Useless Information

ISBN 978-0-06-113405-0 (hardcover)

Visit www.AuthorTracker.com
for exclusive information on your favorite HarperCollins authors.

Available wherever books are sold, or call 1-800-331-3761 to order.